PANIOLO
JOSEPH BRENNAN

Classic head of an old-time *paniolo* Photograph by Herbert Bauer

PANIOLO
JOSEPH BRENNAN

KU PA'A PUBLISHING INCORPORATED
(formerly Topgallant Publishing Co., Ltd.)
Honolulu, Hawaii

Copyright © 1978, 1995 by Joseph Brennan

All rights reserved. No part of this book may be reproduced or transmitted in any form or by any means, electronic or mechanical, including photocopying, recording or by any information storage and retrieval system, without the permission in writing form the publisher.

Second Printing 1995
First Printing 1978

KU PA'A PUBLISHING INCORPORATED
(formerly Topgallant Publishing Co., Ltd.)
3180 Pacific Heights Road
Honolulu, Hawaii 96813
Printed in the United States of America

Library of Congress Cataloging in Publication applied for:
Brennan, Joseph
 Paniolo
1. Paniolo —
ISBN 0-914916-39-4

CONTENTS

Foreword i
Acknowledgments iii
Chapter I Who and What He Is 1
Chapter II The First Sandwich Island Cattle .. 5
Chapter III Delivery and a Ten-Year Tabu 15
Chapter IV A Horned Frankenstein 25
Chapter V The First Horses 31
Chapter VI Unique Cow Country 39
Chapter VII The Early Bullock Hunters 43
Chapter VIII Beef at a Premium 47
Chapter IX The Advent of Vaqueros 51
Chapter X Lassos 55
Chapter XI Saddles 57
Chapter XII Horses, Terrain and Horsemanship 61
Chapter XIII Happy Integration 69
Chapter XIV A New Breed of Man 73
Chapter XV The *Paniolo's* Lot 83
Chapter XVI Seagoing *Paniolos* 87
Chapter XVII The Modern Assembly Line 93
Chapter XVIII The *Paniolos* Proliferate 97
Chapter XIX Robert G. "Boy" Von Tempsky ... 103
Chapter XX "Duke"—The Wild Black Bull 109
Chapter XXI Cattleland Courage 117
Chapter XXII *Paniolo's* Cowtown 125
Chapter XXIII Frank Bernard Vierra 129
Chapter XXIV Joe Pacheco 131
Chapter XXV John Lekelesa 133
Chapter XXVI William Kawai 135
Chapter XXVII Frank Kawai 137
Chapter XXVIII George Purdy 141

Chapter	XXIX	David "Hogan" Kauwe	145
Chapter	XXX	Willie Kaniho	149
Chapter	XXXI	Harry Kawai	153
Chapter	XXXII	A *Paniolo* Salute	159

For my friend Robert Gordon "Boy" Von Tempsky

FOREWORD

In telling the story of the *paniolos* (cowboys) of Hawaii, I hope to spell out special chapters in the mosaic of American history that will serve as valuable and interesting data for those who might be searching for the beginnings of things. I just hope I can create enough undertow to carry the reader along. This will be the history of the beginning of the cattle industry in our Fiftieth State, along with the beginning of the *paniolos* in these far-flung Isles. But it should convey another message, too. It should forcibly remind readers that however troubled the American present may be, its foundation in this country's past was powerfully and bravely wrought and will not easily be brought down. The Island cattlemen are not sorcerers in financial matters, but they use magic on cattle. Every year, with few exceptions, seems to be a vintage year for the ranches.

The *paniolos*, as I see them, are the backbone of the cattle industry in Hawaii, and, as such, represent that derring-do imagination, honesty, character, and hard work that has gone into the building of this democratic nation. In spirit, they're as wild and free as a gull. It's long past time that the *paniolos* should be immortalized in print

— and maybe this documentary will be just one slight contribution in that direction. If so, it will be well worth all the research that has gone into it. For some, it might even rouse a few enduring echoes in the heart...

ACKNOWLEDGMENTS

In researching this book in depth, the author has many people and sources to thank for the unstinting help he was given. Chiefly among them would be: The libraries of both Island newspapers, the *Honolulu Advertiser* and the *Honolulu Star-Bulletin*, the Hawaii State Archives, the Hawaii State Library, the University of Hawaii Library, Richard Smart, owner of the Parker Ranch, old issues of the long defunct *Paradise of the Pacific Magazine*, the late Emma Lyons Doyle, volumes of Captain George Vancouver's log, *A Voyage of Discovery to the North Pacific Ocean and Round the World* (N. Israeil, Amsterdam Da Capo Press) New York 1967, Robert G. von Tempsky, cattleman of Maui, the *Honolulu Magazine*, Larry Kimura's *Old-Time Parker Ranch Cowboys* (Hawaii Historical Review, 1964), Richard K. Kimball, and many old-time Island *paniolos* named herein, who were gracious in answering my many, many questions. Many of these sources provided the necessary photographs used in PANIOLO. There photographs aid in giving an accurate account of the life of the *paniolo*.

Without the assistance of these varied sources, this documentary might have wound up shy on much of its authenticity and factual datum. The author has them to thank, and hereby gives them much, much *mahalo* for their help.

I

WHO AND WHAT HE IS

The song of the colorful cowboy of America's great Southwest has been sung so well, so often, and so long that one is daunted at even trying to describe a kindred brand of that breed of man. We already know the American cowboy who is so adept at cutting and gathering the cattle herds on the western prairies and working them forward to paddocks, to railroad loading stations, to brandings, or maybe even to better grass on distant pasturage. He herds them away from flashflood regions and swirling waters, he pushes them away from vast grass fires, he shoves them out of snowbound regions and into areas where feed is available, and he turns them and calms them when they stampede from storm fright.

Observe him at these chores sometime and you'll know you're watching an all-out, go-for-broke performance. In all these things he is a solid craftsman and a dedicated one, whether it involves the cutoff of the sudden breakaway dart of a snuffy steer, or the magic lassoing of a renegade bull bent on destruction of some crazy kind. Him we know. With rope, horse or steer, he has the magical touch.

But there's another cowboy who rides other plains —

far distant plains in an almost improbable place — who is a by-product derived from that very old Southwestern source. He is as colorful, as brave, and equally as skillful as those men who have been riding the cattle ranges in the Southwest for so many generations. This Island rider has made the huge cattle kingdoms of Hawaii what they are today. And, believe it, Hawaii's cattle history involves years of tumult and disaster, triumph and ineptitude and daring. Seldom does a cattle ranch have a brush-fire success. The *paniolo* has ridden his way through the toughest and the best of it — and he has learned to shrug a lot in his herding. Still, he generally winds up with that look of joy unconfined.

He's a hard riding man, leather-tough — a laughing man — and often with a lei of pansies around the crown of his hat. He has what the Hawaiians call *manao* — the spirit and feel of the true Hawaiian. Some say he is a lonely man; lonely because he rides alone so much. Maybe the quiet hills and plateaus bring on that quality. We see the *paniolo* as a person with the quiet sensitivity of a man who has lived for a long time with his loneliness and who is now on good terms with it. He's a man who understands the whispered language of nature. He is also a quality cowboy.

Yet, exactly who is this little known rider to whom we refer? None other than the *paniolo* of Hawaii. Let us tell you about him; he cuts quite a figure as a cowboy — and he's in the Islands to stay. Reason? Surprisingly, hordes of cattle tread the Islands, particularly on Hawaii and Maui, and somebody has to be their keeper. The Big Island of Hawaii alone has one of the largest spreads in the world, and its hundreds of *paniolos* over the years have left a legendary history. And they still ride today. They ride Hawaii's mind-blowing mountain, valley and prairie terrain — and make the magnificent most of it.

The *paniolos* we've known look at each newborn day as something special. They saddle up their horses and look out at the majesty of the land they will ride, and they know the day will have its hazards. But they seem to toss the world over their shoulder for good luck. They ride away in early morning and it's as though it's a new, exciting life they're beginning; it's the parachutist's jump, the eagle's dive, the search for the Holy Grail and for the white whale. Before the day is out they might be straining almost to collapse, but right now they're smiling crisply. Their daily road is not rose-strewn. They might not be back until the moon rides at anchor in the sky, but each returning *paniolo* will be a man at peace within himself.

With the *paniolos*, living is a horse between your knees, a range of brooding mountains in the distance, a day of clouds with the sun painting a miracle of gold and blue in the heavens, and a herd of lowing cattle to tend. Give them these things and they are a success at every level — and watching them is a solid joy. Maybe it's because, the year round, they have April in their hearts. April because they love what they do.

Every *paniolo* has seen many fine and stirring things: his wife at the altar, his children reaching up to him for affection, star-invaded nights with every star a gem of purest rhinestone, and morning fogs wine-colored with dawn. But nothing he knows surpasses the sight of a lush-green Hawaiian valley or hillside dotted with sleek red cattle grazing in the sun. It is his opium dream come true.

The original *paniolos* were Spanish-Mexican vaqueros imported from California to the Islands for the purpose of working with the Hawaiians and teaching them how to handle the great herds of wild cattle and horses that ranged the slopes and valleys of Hawaii. The Island natives were soon Hawaiianizing the word *Espanol* to *Paniolo*, and the name still has currency to this day. The word is spoken with well-earned respect and admiration, for the Hawaiian cowboy of today has truly stamped his mark of worth on the Island's image. He is rated very tall in his craft.

It stands to reason that the vaquero had to have something tangible and promising to come to — something worthwhile that would encourage him to leave his sunny California and its friendly range country. He already worked under skies colored like a pigeon's neck; he already loved his moon-drenched prairies and hills. This new country would have to have cattle, there would have to be horses, and there would have to be rangeland — the things he knew, loved and worked with so well. Rangeland there must be, for that was his workbench. Horses? They were his tools. Cattle? They were his product. They made up the world of the vaquero, and he needed that world; it was his sun and air.

So, first things first, our story must necessarily first deal with how the cattle were obtained and shipped to the Hawaiian Islands (the then Sandwich Islands), and how the first horses were imported and turned loose there. In short, had there been no cattle and no horses, in the Islands, there would have been no employment for these lusty, bronzed centaurs of the range.

II

THE FIRST SANDWICH ISLAND CATTLE

It all really began when the British Captain George Vancouver picked up some cattle in January of 1793 at the Spanish mission of Monterey, California. Of course, one might wax even a little more pragmatic and insist that the beginning actually dates back to 1521 when the Spaniards sailed to Vera Cruz, Mexico and brought the first cattle to North America. Those conquistadors had assumed that the animals carrying the blood lines of their fighting bulls of the arenas would be hardy enough to survive in Mexico's wilds. The Spaniards were right, for the beasts throve and multiplied into countless numbers.

In any event, 272 years later at California, Captain Vancouver took some of their descendants aboard his sloop of war, HMS *Discovery*, for delivery to King Kamehameha the Great in the Sandwich Islands. The English explorer intended to demonstrate that his visit with the Polynesian king was in the interest of friendship, and he felt that the gift of the cattle would help prove his point. His journal bears out that he wanted to contribute to the welfare of the Sandwich Islanders, for he wrote in this manner of his departure from Monterey:

"Senor Quadra's (Spanish commanding officer at that

port) benevolent disposition encouraged me again to obtrude on his goodness by requesting some black cattle and sheep, for the purpose of establishing a breed of those valuable animals in the Sandwich Islands. A dozen, being as many as we could possibly take on board, were immediately provided, consisting of four cows, four ewes, two bulls, and two rams. The prospect we had of a good passage to those islands induced me to lay myself under this additional obligation, hoping by such an importation, to accomplish at once the purpose I had in contemplation; which, if effected, could not fail of being highly beneficial, not only to the resident inhabitants, but also to all future visitors."

Accompanying the *Discovery* was Vancouver's small consort vessel, the armed tender *Chatham*. The countless leagues of distant water daunted them not. Hardy men crewed the tough little ships. It was a long sail, a hard sail, and the Sandwich Isles were to be but one stopover as part of Captain Vancouver's round-the-world voyage of discovery. One of his purposes was to establish better relations between his own England and the Hawaiian people. Since his previous voyage to the Islands in 1792, the natives had shown much displeasure with visiting foreigners, and actual fighting had erupted where loss of life took place on both sides.

Among Vancouver's other intended projects, he planned to erect temporary observatories on shore to make astronomical observations. He would also be involved in botanical research, plus charting the coastlines and shore waters of distant lands and islands. In short, he would be collecting much valuable data in the interest of science.

En route over the trackless Pacific Ocean wastes, Vancouver strove to keep his cargo of cattle healthy and alive, but it was no easy thing. Cramped space on a pitching sailing ship was no ideal place for animals that are in the habit of ranging widely. Fresh water and green fodder were at a premium. Added to this, of course, was the

The First Sandwich Island Cattle

prospect of meeting with open hostility from the Islanders. Alarming reports of native attacks on visiting ships had filtered back from the Islands, and Vancouver wasn't even sure that his cargo of cattle would be accepted. The Islanders, warring between themselves, had, of late, only been willing to accept guns and ammunition as a form of barter. And Vancouver was under strict orders of His Majesty of Britain to let no such items fall into the hands of the Sandwich Islers. Firearms only worsened the inter and intraisland slaughtering. To bring and promote peace was a prime part of Vancouver's plan.

But it could not be forgotten that just fourteen years before this — 1779, to be exact — that Captain James Cook, the British navigator explorer, had been killed by the natives at Kealakekua Bay on the Big Island. In the interim, other foreigners, too, had been brutally killed at various points in the Islands, consequently, the dangers remained real and present.

So it was that Captain Vancouver brought his two vessels to the northern coast of the larger island, Hawaii (spelled and pronounced O-why-ee by him), and there he parted company with his consort, sending the *Chatham* south along the east coast, while he pursued the southerly route down the west coast. The plan was for both ships to survey the coasts for harbors, then to rendezvous at Kealakekua to the south.

The *Discovery* moved on under full sail, the tradewinds stiffening and a white wake veeing out aft of her. Vancouver and his officers continued scanning the shores with their glasses, studying what they could see. The shoreline, for the most part, looked bleak and unwelcome; tumbling breakers on the rocky shores, tall, sloping mountains in the background — and certainly no sight of tilled soil or habitations. The inhabitants themselves were noticeable only for their total absence, and the captain resumed wondering about his welcome — or lack of it . . .

So, here in 1790 Vancouver was off Hawaii's shore —

and time was running out on the cattle. They certainly had not weathered the long, rough voyage very well and were more than a little sick. They had undergone too many days and nights of water shortage and lack of green fodder. Too many blazing days aboard ship had been like days under a glass bell. With a brassy sun scorching down from cloudless skies and nary a breeze stirring on some days, there had been nothing but glare and hot, parched air for the cattle to suffer through. The ship's live cargo looked like Disaster, Incorporated. The chances were that none of the animals would survive the balance of the trip.

It was midmorning when a native canoe finally put off from one of the beaches and approached the *Discovery*. Vancouver brought his vessel about in order to rendezvous with the canoe's occupants. All alarm aboard the *Discovery* subsided at sight of so small a delegation of bronzed natives, for it seemed to be an omen of good things to come. No weapons were in sight aboard the canoe, and no additional craft were bringing up the rear. These were black-haired, well-muscled men, barefooted and wearing *malos* (breechclouts) as their only apparel.

The natives soon gave Vancouver to understand that King Kamehameha was farther to the south at Karakakooa (Kealakekua), and that a general tabu prevented other natives from coming off shore. They also advised that, under penalty of death, they were prevented from trading hogs, vegetables, etc. for anything other than arms and ammunition. They did, however, despite the tabu, trade some of their cargo in the way of one hog, two or three fowl, some vegetables and breadfruit; all in exchange for some pieces of iron.

The canoe returned to shore, and Vancouver continued sailing his vessel southward along the coast. Although he loafed close to shore, no other canoes approached, nor did natives make an appearance on the cliffs or on the beaches. He began to be apprehensive about the alleged tabu, suspecting that possibly there was really some serious or ominous reason for the nonappearance of natives either on shore or on the water. Maybe it had to do with some kind of planned surprise attack...

The sloop of war edged farther along shore, glasses still trained on its contour for miles. Finally, habitations were sighted as the vessel moved farther southward. Ultimately another canoe made its way from shore and came alongside. Once more the story of the tabu was told. This group of natives said they belonged to a chief by the name of Kahowmotoo, and that the latter could be found residing farther south at a village named Toeaigh (Kawaihae). They explained that the rank and consequence of their master, Kahowmotoo, authorized him to dispense with the tabu restrictions on the present occasion. The ship moved on, and there was still room for speculation as to what kind of reception would be given these foreign interlopers.

So, this was the dilemma of the *Discovery's* captain and crew when the ship nosed into Kawaihae Bay on the 14th of February, 1793. Anything could happen. The village might well harbor armed Polynesians eager to destroy any and all outsiders. The prospects of it had everyone aboard the sloop as tense as a coiled spring.

The ship dropped anchor and soon canoes were seen coming out from shore. To Vancouver's comfort, there appeared to be no evidence of hostility. To the contrary, the paddlers brought high chief Kahowmotoo alongside with a gift of six hogs and a supply of vegetables. Vancouver reciprocated with a gift of a ram, two ewes, and a ewe lamb that had been born on the ship's passage.

Although the chief was disgruntled over not being able to get any firearms or ammunition in return from the *Discovery* (again due to the restriction laid down by King George of England), he seemed pleased with the exchange, and strongly urged that Vancouver remain a few days there at Kawaihae. To the *Discovery's* captain and men, these meager signs of friendship were as welcome as a footprint on Crusoe's island.

Captain Vancouver accepted, for there were the prospects of picking up supplies — particularly some fresh feed for the now half-starved cattle. He wrote in his log:

"To these (Kahowmotoo's) solicitations I in some measure consented, by promising to stay the next day in the expectation of not only deriving some supplies for ourselves, but of procuring some provender for the cattle and sheep; which, in consequence of the inferior quality of the hay obtained at Monterey, were almost starved. To this cause I attributed the unfortunate losses we had sustained in our passage, amounting to three rams, two ewes, a bull and a cow. These were serious misfortunes, and in a great measure disappointed the hopes I had entertained, from the importation of these valuable animals into the several islands of the Pacific Ocean. Still, however, I flattered myself with the expectation of succeeding in Owhyhee (Hawaii), by leaving the remaining bull, with the rest of the cows, under the protection of Tamaahmaah (Kamehameha), who I expected would meet me at Karakakooa (Kealakekua), to receive and insure as far as possible, the preservation of the animals I had on board."

Meanwhile, the chief again reassured Vancouver that Kamehameha was currently at Kealakekua, some thirty miles farther south on the Kona Coast. Eyeing the remaining cattle aboard the *Discovery*, and learning that they were for the king, he offered to take them into shore and hold them in safety for Kamehameha's return to Kawaihae. Stressing what Vancouver already knew, he insisted that good pasture would soon restore the animals to excellent health. The captain was tempted to release them at once, but after more sober consideration, he thought better of it and resolved to present the animals direct and in person to the monarch. In addition, Vancouver was a little skeptical of the chief's professed friendship and loyalty to Kamehameha.

Just the same, the captain remained over in the harbor and did some visiting and trading with the chief. No arms or ammunition were used in bartering. Instead, upon receiving another sixteen very fine hogs and an assortment of fresh vegetables, Vancouver repaid the chief with two yards of red cloth, a small piece of printed linen, plus some beads and other trivial articles. The latter items were for the chief's four wives, who still remained ashore due to the prevailing tabu which prohibited women from embarking in canoes. The few girls who actually came aboard the *Discovery* had swum the whole distance.

During the stopover Vancouver went ashore and visited the village. He found it to be situated in a grove of tall coconut trees just behind a sandy beach. Caution was still the order of the day, for he was accompanied by a corporal and six armed marines. He found the village consisting only of straggling grass houses of two kinds; the first type were the miserable living quarters which amounted to nothing more than huts; the second kind were the structures used for building, repairing and storing canoes.

Captain Vancouver examined their crude mud-and-stone reservoirs which were set up for the obtaining of salt deposited by the incoming seas. He paid his respects to the chief's several wives, gave away additional wanted articles and trinkets — and found no hostility from anyone. He even had another chief — Tianna by name — come aboard with another dozen fine hogs, and there seemed no end to the hospitality extended by these Sandwich Islanders.

Two days later, the 16th of February, 1793, despite the sickness of the cattle and the offer of the high chief to care for the animals, Vancouver took advantage of rising winds and set sail for Kealakekua to the south. It was a windswept, rainy Gothic day.

Unfortunately, he was soon heading into violent weather. Heavy seas began boiling over his bow, big winds wailed through his rigging. The *Discovery* lurched stubbornly on her southerly course until just the reverse kind of weather set in — a total calm. It lasted and it lasted. Towards dark, with breezes finally freshening, they sighted a brig and a sloop on the distant horizon. Through their glasses they saw that the first was the consort ship, the *Chatham*, the armed tender which had parted company from the *Discovery* at the northern tip of the island; the second vessel was the *Jackall,* a trader. The *Discovery* strove to overtake them but lost them in the night when subsiding winds once more left her becalmed. For a long while she lay dead in the sea.

The morning of the 18th again found the vessels within sight, and by noontime the *Discovery* was able to rendezvous with the *Chatham* and exchange much wanted information. Captain Vancouver was forewarned that King Kamehameha had procured some cannons and ammunition from certain traders and the guns were mounted on stonewall emplacements at Kealakekua. The skipper didn't relish the news, for the prospects of facing land guns tightened the scalp. Nor was he happy about all the delays he had encountered of late. Aimless drifting, protracted calms, the extended layover at Kawaihae, plus intermittent baffling winds, had thrown him way off schedule. Meanwhile, his remaining cattle were even more visibly failing in health. It didn't appear that he would even get them to the king while there was still life in them. Time was wasting. Captain Vancouver referred to the delay with real dismay when he wrote:

"A circumstance now occured that contributed to make me infinitely more dissatisfied with this irksome detention from the shore. The only bull that remained, and a cow that had brought forth a dead calf, were no longer able to stand on their legs, and it was evident that if a speedy opportunity did not offer itself for relieving them by sending them on shore, their lives could not possible be preserved. The loss, particularly of the bull, would have been a cruel disappointment to my wishes; but as favorable circumstances often take place when least expected, so it was on this occasion."

The "Favorable circumstance" came in the form of a

visit from the half brother of King Kamehameha, Chief Crymamahoo. The chief was accompanied by a flotilla of canoes and, despite the fact that they were now some twenty-five nautical miles from shore, Vancouver immediately sought to have the two sickest cattle taken ashore in one of the larger double canoes. At first the chief would have no part in transferring the animals to land, and he was more than a little vociferous in his refusal. Not until Vancouver offered substantial payment did the chief finally agree to assist.

III

DELIVERY AND A TEN-YEAR TABU

Captain Vancouver's journal of February 19, 1792 reads: "I offered Crymamahoo (well knowing that avarice is a predominant passion with many of these Islanders) a moderate recompense only, for allowing his canoe to perform this service. He instantly waved all his former objections, and the bull and cow were soon comfortably placed in his canoe, in which there were some vegetables that the bull ate, seemingly with much appetite; this gave me great pleasure, as I was now in hopes that he would soon recover by the help of proper nourishing food which the shore abundantly supplied."

The *Discovery* separated from the canoes and continued her voyage southward under full sail. Word about the guns to the south made the suspense aboard ship as taut as a tow line. Again the sloop was met by additional large and small canoes laden with various fresh supplies. This time Kamehameha's eldest son was aboard one of the outriggers. The lad was approximately nine years old, bright, pleasant, and the one slated to someday succeed his father as king.

Vancouver was tremendously impressed with the young prince. There was much exchange of gifts. The natives were particularly grateful to receive red and blue woolen cloths and printed linens. Beads and other trinkets were accepted, too, but were held in much less esteem.

They parted company and again the *Discovery* sailed southward, this time under light breezes. Due to the threat of the cannon emplacements reported to be at Kealakekua, Captain Vancouver decided — according to his journal — to first transact his business at Tyahtatooa (Kailua-Kona) before facing the "inconvenience" of possible heavy cannonading at Kealakekua. Arriving offshore from Kailua-Kona, he dispatched his officer, Whidbey, in the ship's cutter to examine the anchorage. Then, what with contrary winds and heavy crosscurrents, the *Discovery* was driven a considerable distance from shore, and this naturally contributed to more unwanted delay.

At noon Vancouver was surprised to find more canoes approaching him — this time actually with King Kamehameha and some of his picked warriors and advisors in the vanguard. It seemed to augur well for the kind of reception the *Discovery* would receive. Yet, one couldn't tell for sure . . .

Kamehameha, in his impressive and austere manner, rode in the lead canoe, and his imperial raiment was that of a full-length feathered cloak and a high, arched feathered helmet to match. The powerful body beneath the golden cloak made him resemble a great bird of prey on the verge of spreading its wings and flying with deadly effect at any adversary. Vancouver eyed the oncoming flotilla with mixed awe and misgivings, but on closer scrutiny he saw several women aboard some of the canoes. He was convinced that this was merely a peace mission.

The lead canoe came alongside the *Discovery* and Vancouver got a better and closer look at the king. Apropos of this, he wrote in his journal: "Not only from Captain King's description, but also from my own memory (Vancouver had first seen Kamehameha in 1779), as far as it would serve me, I expected to have recognized my former acquaintance by the most savage countenance we had hitherto seen amongst these people; but I was agreeably surprised in finding that his riper years had softened that stern ferocity which his younger days had exhibited, and had changed his general deportment to an address characteristic of an open, cheerful, and sensible mind; combined with great generosity, and goodness of disposition. An alteration not unlike that I have before had occasion to notice in the character of Pomurrey at Otaheite (Tahiti).

"Tamaahmaah (Kamehameha) came on board in a very large canoe, accompanied by John Young, an English seaman, who appeared to be not only a great favourite, but to possess no small degree of influence with this great chief. Terrehooa, who had been sent to deliver the bull and cow to the king, was also of the party, and informed me that the cow had died in her passage to the island, but that the bull arrived safe, and was lodged in a house where he ate and drank heartily."

Upon Kamehameha's request, Captain Vancouver also took aboard the king's queen in addition to many of the monarch's friends and relatives. More and more native canoes paddled up onto the scene, but Vancouver stuck to his safety measure of allowing only the principal chiefs aboard. It lessened the chance of a native take-over of the ship, as had been done several times in the past. The sloop's skipper distributed many gifts to the visitors, and they seemed exceedingly pleased with each present. The occasion developed into a time of gayety. When the king was gifted with a scarlet cloak that reached from his neck to the ground, adorned with tinsel lace and blue ribbons to tie it down the front, he entered into the spirit of the thing in a way that promised real success for the *Discovery's* whole visit.

As night fell the natives took their leave of the sloop and paddled away in their canoes. Meanwhile, all arrangements had been made for the *Discovery* to drop anchor at Kealakekua where the king kept his residence.

Vancouver had learned that the rumors of fortifications and cannons at Kealakekua were entirely false.

The vessel set sail for the southern port and didn't come abreast of it until early morning the following day, the 22nd of February, 1793. Immediately a vast squadron of native canoes met them offshore all prepared to exchange merchandise and, hopefully, come aboard. Vancouver restricted all bartering until his vessel could be properly anchored and secured close into the beach. The rowdy weather was behind them now, and they were enjoying picture postcard perfect weather. The hypnotic scenery ashore beckoned them; the blue-blue bay and its encircling coconut palms waving their green fronds in welcome.

Before the anchors had even been dropped, eleven large additional canoes set off from shore in a formation that told of a special visit. They formed two equal sides of an obtuse triangle, the biggest one came on at the fore part of the triangle with eighteen paddles working on each side of the craft. And, sure enough, there in the center was his majesty, King Kamehameha, dressed in a printed linen gown with a most elegant yellow-feathered cloak over his massive shoulders. Again his head was helmeted by another feathered thing of beauty that gave him a savage regality. He was all stateliness and the thousands of feathers shouted of the vast bird life that had made possible the helmet and full-length cape.

The silent procession came forward with a precision of paddling that was regular as clockwork; solemn and impressive. All of this was carefully and magnificently done under the king's direction. The *Discovery's* officers and crew stood in awe over the precision of the maneuvers as the flotilla circled the sloop. There was dignity and majesty in it. Suddenly Kamehameha ordered his following ten canoes to close in and remain astern of the *Discovery*, while his lead canoe swept to the sloop's starboard side and halted.

Kamehameha immediately ascended the side of the

Delivery and a Ten-Year Tabu

ship and took the hand of Captain Vancouver. He demanded to know if the ship's skipper was his sincere friend, and when assured this was the case, he wanted to know if King George of England was likewise his friend. Upon being reassured of this also, he instantly touched noses with Vancouver as a token of trust.

There followed an exchange of compliments and gifts, and the monarch announced that he had ninety very large hogs in the canoes for the *Discovery*. These latter, Vancouver tried to decline, for his vessel was already more than a little crowded; but the Hawaiian king would not take *no* for an answer.

Vancouver now had his opportunity to present the balance of his cattle to the monarch. The captain wrote in his journal on February 22, 1793:

"The remaining livestock I had on board, consisting of five cows, two ewes and a ram, were sent on shore in some of his canoes; these were all in a healthy state though in low condition, and as I flattered myself the bull (at Kawaihae) would recover, I had little doubt of their succeeding to the utmost of my wishes. I cannot avoid mentioning the pleasure I received in the particular attention paid by Tamaahmaah (Kamehameha) to the placing of these animals in the canoes. This business was principally done by himself; after which he gave the strictest injunctions to his people who had the charge of them, to pay implicit obedience to the directions of our butcher, who was sent to attend their landing."

The landing of these cattle, and the antics of the land-starved cows were something to witness. The natives, having never before seen such massive animals, watched the horned beasts cavorting and stampeding for everything green in sight. These strange animals from a foreign land staggered the minds of the Hawaiians, and they ogled them as something out of a bad nightmare.

On advice of Captain Vancouver, Kamehameha put a ten-year tabu on the cattle — stipulating that they were not to be molested or killed on pain of death. Commoners and *alii* (of royal blood) alike well knew that the king's threat of punishment would be carried out should anyone break his law. So, barring disease or some holocaust of nature, there appeared every reason to believe that the animals would multiply greatly in years to come. The order was that the animals would be allowed to roam unmolested at large over the island's mountain slopes and prairies and in the valleys and wildernesses.

One of the chiefs, Kahowmotoo, was plainly irritated over the division of livestock. In private he gave Vancouver to understand that he was glad the cattle and sheep were being introduced to the island, but he thought the distribution was unequal and unfair. His complaint was that all the large cattle had been unjustly given to Kamehameha. Vancouver had to point out that an injustice had already been done to the king when he, Vancouver, had given Kahowmotoo the sheep originally designed for the king. Ultimately, the captain sought to appease the chief by promising to him a supply of cattle on his next trip. It seemed to placate the disgruntled chief.

Then the other chief, Tianna, showed up and he, too, displayed a disposition to be ungrateful and disagreeable. All this spleen could well lead to complications, and Vancouver knew it. Nevertheless, he elected to pay his principal court to Kamehameha, the king of the whole island, and trust to the lesser *alii* settling for smaller attention.

Hordes of natives continued to multiply on the shore and in the waters. According to Vancouver's estimate, some 3000 of both sexes were in canoes or swimming about the sloop. He wrote that their clamor was stunning. The warm metallic sea seemed bouncing with people and canoes, and, so far, everything portended to be hail and farewell between the ship's crew and the natives.

The following day, however, was a day of peace and quiet without a single canoe visiting the *Discovery*.

Then there ensued several days of more negotiations

with Kamehameha. And there was heavy work to do; setting up a camp and observatory on shore, repairing the ship's defective mast, tending to her damaged rigging and sails, and taking on additional needed supplies of water, wood and vegetables.

Vancouver visited shore with Kamehameha and was impressed with the efficiency and astuteness of the king's advisor, Young, who was outstandingly co-operative. Still cautious, however, the captain took a marine guard of six ashore with him and saw to it that his field pieces were mounted in readiness on the ship's quarterdeck. Too, every crewman was alerted to the possibility of being called to battle stations on an instant's notice. Vancouver knew the insatiable desire these natives had for pistols and muskets, and he realized that this one weakness of their could erupt into violence at any moment. He had already witnessed Islanders wrenching firearms from the hands of officers and midshipmen, and the danger was everpresent.

But Kamehameha and his people continued showing friendliness throughout. The king even applauded Vancouver for his precautions, pointing out that there were certain chiefs and people who just might prove hostile when least expected.

On the 27th of this month, the king came up with some bad news which measurably dampened Vancouver's spirits. The captain made this entry about it in his journal:

"He (Kamehameha) gave me the unwelcome intelligence that the bull for whose recovery I was so very solicitous, was dead. On this mortifying occasion I much regretted that I had not followed the advice of Kahowmotoo, from whose connection with the king I most probably might have relied with perfect security on his offers, of taking charge of the cattle at Toeaigh (Kawaihae). Two of the young cows, however, appeared to be in calf; this encouraged me to hope that his loss would be repaired by one of them bringing forth a male. The finest of the two ewes, I was now informed was killed by a dog the day after the cattle were landed; whose life was instantly forfeited for the transgression."

Well, that left but five cattle remaining alive on the island — and all of them cows! Captain Vancouver had solid reason to worry over his cattle-importing venture; the chances were that neither one of those calving cows would bring forth a bull calf.

However, history shows that Captain Vancouver made sure of this original cattle venture becoming a success by returning to Hawaii again the following year and bringing additional cattle. Evidence that he picked up at least one additional bull is found in his journal dated Nov. 20, 1793. He wrote:

"When we reached the shore (mission at Buena Ventura, California) the surf still ran very high, but with the assistance of our light small boat we landed with great ease, perfectly dry, and much to the satisfaction of our worthy companion (Friar Vincente Santa Maria); of whose bounty there was yet remaining near the beach a large quantity of roots, vegetables, and other useful articles, with five head of cattle, in readiness to be sent on board. One of these being a very fine young bull was taken on board alive, for the purpose of being carried if possible to Owhyee. The others were killed, and produced us an ample supply; had they not been sufficient, a greater number were at hand, and equally at our disposal."

Then, upon arrival again in the Islands, he wrote in his journal, dated January 15, 1794;

"After the large canoes had delivered their acceptable cargoes, they received and took to the shore (Kealakekua Bay) the live cattle, which I had been more successful in bringing from New Albion (California) than on the former occasion. These consisted of a young bull nearly full grown, two fine cows, and two very fine bull calves, all in high condition; as likewise five rams and five ewe sheep. Two of each of these, with most of the black cattle, were given to the king, and as those I had brought last year had thrived exceedingly well; the sheep having bred, and one of the cows having brought forth a cow calf; I had little doubt, by this second importation, of having at length effected the very desirable object of establishing in this island a breed of those valuable animals."

So it was that cattle were brought from North America to these Sandwich Islands. History now bears out how they multiplied and thrived beyond all belief. But one of the biggest contributing factors to their increase in numbers is due to that tabu arrangement which Captain Vancouver made with King Kamehameha; he got that ironclad promise out of the monarch which assured the propagation of the cattle.

Vancouver wrote in his journal on February 23, 1794: "Anxious lest the object I had so long had in view should hereafter be defeated; namely, that of establishing a breed of sheep, cattle, and other European animals in these islands, which with so much difficulty, trouble, and concern, I had at length succeeded so far as to import in good health and in a thriving condition; I demanded, that they should be tabued for ten years, with a discretionary power in the king alone to appropriate a certain number of the males of each species, in case that sex became predominant, to the use of his own table; but that in so doing the women should not be precluded partaking of them, as the intention of their being brought to the island was for the general use and benefit of every inhabitant of both sexes, as soon as their numbers should be sufficiently increased to allow of a general distribution amongst the people. This was unanimously approved of, and faithfully promised to be observed with one exception only; that with respect to the meat of these several animals, the women were to be put on the same footings as with their dogs and fowls; they were to be allowed to eat of them, but not of the identical animal that men had partaken, or of which they were to partake."

IV

A HORNED FRANKENSTEIN

It must be understood that this was Captain Vancouver's third visit to the Sandwich Islands — originally under Captain Cook's command, then again in 1793 with his own command — and he had won the confidence of King Kamehameha and most of the chiefs and commoners. They believed in him and were quick to accept his suggestions, like the tabu plan for cattle. In truth, this third trip was made basically to conclude the ceding of the island of Hawaii to the British crown. Kamehameha had already agreed to it, for he wanted England's protection from other countries, but it still remained to working out the final ceremonies which would make it official.

Apropos of "protection," King Kamehameha had enumerated the several foreign nations that, since Captain Cook's discovery of the islands in 1778, had taken advantage of the Sandwich Islanders. The foreign powers had been too powerful for the Islanders to resist, and Kamehameha now wanted his favorite civilized power, Britain, to protect his tiny, defenseless kingdom from any foreign encroachments.

So now much was going on here at Kealakekua Bay in 1794 in the way of further negotiations. Speeches were being made, agreements discussed, papers signed; then, as Captain Vancouver wrote: "All assembled on board the *Discovery* for the purpose of formally ceding and surrendering the island of Owhyhee to me for His Brittanic Majesty, his heirs and successors."

When all arrangements for cession were satisfactorily made and agreed upon, the news was immediately made known to the natives in the flotilla of canoes surrounding the sloop. The Hawaiians shouted their approval, cheered and passed the word along toward shore. Captain Vancouver wrote: "Mr. Puget, accompanied by some of the officers, immediately went on shore; there displayed the British colours, and took possession of the island in His Majesty's name, in conformity to the inclinations and desire of Kamehameha and his subjects."

There was no reason to be further detained in Kealakekua, so on February 26, 1794, Vancouver sailed his ship northward to Tyahtatooa (Kailua-Kona) where he again dropped anchor. He was accompanied by King Kamehameha and some of the latter's chiefs. They visited the royal residence at this place. Thence they sailed farther north again to Toeaigh (Kawaihae) where, the year before, Vancouver had put ashore the first two head of cattle.

It is noteworthy that Vancouver thought well of this region as an excellent area for the propagation of the cattle which he had left on the island. He wrote: "The country rises rather quickly from the seaside and, so far as it could be seen on our approach, had no very promising aspect; it forms a kind of glacis, or inclined plane in front of the mountains, immediately behind which the plains of Whymea (Waimea) are stated to commence, which are reputed to be very rich and productive, occupying a space of several miles in extent, and winding at the foot of these three lofty mountains far into the country. In this valley is a great tract of luxuriant, natural pasture, whither all the cattle and sheep imported by me were to be driven, there to roam unrestrained, to 'increase and multiply' far from the sight of strangers, and consequently less likely to tempt the inhabitants to violate the sacred promise they had made; the observance of which, for the time stipulated in their interdiction, cannot fail to render the extirpation of these animals a task not easily to be accomplished."

Time proved Captain Vancouver to be so right, for the thousands of cattle that later swarmed over valley, slope and dale in this very region bore mute testimony to the land's rich, natural pasturage. They throve and grew wild in vast herds, and became great destroyers of the natives' crops. The Hawaiians showed a towering respect for the agility and sheer brute strength of the beasts. Many people were attacked, many hurt, some killed — and the hairbreadth escapes from onrushing horns were common enough.

Something of a Frankenstein had been created. In fact, the wild cattle became so belligerent and hostile that they even represented a menace to travelers. Many visitors to the Island were attacked and many hurt. One notorious instance which is still talked about involved the world renowned botanist, David Douglas, for whom the Douglas fir was named. The wild cattle were so hazardous to approach that the bullock hunters turned to digging deep pits to trap the beasts. David Douglas's mutilated body was found in such a pit in the Waimea area of Hawaii in 1834. The great Scotsman was on a plant-collecting excursion when his end came, and to this day it still remains a moot question as to exactly how he came to his violent death.

For a long time the supposition had been that Douglas accidentally fell into the pit-trap and was killed there; either the already trapped bull had gored and trampled him to death, or the bull had fallen in upon him. But suspicion was rampant because a substantial amount of gold was missing from the dead man's pockets. Reverend Lorenzo Lyons told of the sad-faced Hawaiian who brought in the corpse to Douglas's friends, "with the dreadful intelligence that the body had been found in a pit excavated for the purpose of taking wild cattle, supposed to have been killed by the bullock which was in the pit."

Douglas's friends were numbed by the sight of the mutilated body. The man's ribs were broken, clothing torn, great gashes on the head. Mournfully, they prepared for the funeral and he was properly buried. But suspicion still lurked, for at the inquest a suspected well-known cattle hunter — a Botany Bay ex-convict, by the way — gave testimony, and the deposition read this way in part:

"He states (Goodrich and Diell to English consul) that on the 12th inst. before six in the morning, Mr. Douglas arrived at his home, asked him to point out the road and go a short distance with him. After taking breakfast, the Englishman accompanied Mr. D. about a mile and a half. He warned him particularly of the three bullock traps, two of them directly in the road . . . About eleven o'clock the natives came saying the European was dead . . . At the pit he (the Englishman) found the bullock standing on the body."

Mystery still shrouded the disappearance of the gold which Douglas was known to have carried on his person. Just the same, medical men had made their examination and were of the opinion that the wounds on Douglas had been made by a bullock. So it went.

Yet, many years later a bullock hunter confessed on his deathbed that he had murdered the traveler for his gold, then had thrown Douglas's body into the pit with the wild bullock. It's a historical fact that many of the bullock hunters hired by King Kamehameha III were men from Botany Bay, a colony for criminals in Australia — and this dying man was one of them. It's a truth that an element of men in the Waimea area represented a maggoty and transient world — many of them felons and fugitives from justice.

Another statement made by a fine old Hawaiian of unimpeachable integrity was: "How could the botanist have tumbled into a pit when bullock had already broken through the disguise? Why were wounds only on the man's skull? What became of Douglas's money?" All good, sound questions. This *ahapaha* Hawaiian — Balabola by name — went even further, in fact, and insisted: "No, the *haole* was murdered. We were afraid at the time and only whispered it among ourselves. And when my father died, and old Kilino (another noted Hawaiian), they each repeated the story to me."

In short, the story of the killed Scot botanist merges with the saga of Hawaii cattle. And it seems that cattle and cattle raising in the Islands, and all that the two connote, have always been tied in with drama of some kind or other — sometimes glorious and fine, sometimes tragic and deadly.

In the overall picture, wild cattle had certainly put the early Hawaiians in a dilemma. They had their trials and tribulations with them right from the start. It would take stout and brave men to someday bring the animals to hand and turn them to the good of the Islands and the beef markets in far-flung places. It would take horses, too — sturdy, intelligent horses that could co-operate with the men with the precision of clockwork. Capable men and trainable horses were a must — and they had to come!

V

THE FIRST HORSES

The horses came more than a quarter of a century ahead of the *paniolos*, and 1803 was the year when the animals were first introduced to the Islands. It came about in this way: Captain Richard J. Cleveland, aboard the brig *Lelia Byrd*, sailed from the Bay of San Quentin in California in 1803, and on March 20th of that year picked up a stallion and a mare with foal at San Borgia. He obtained the animals from a Padre Mariano Apolonario of St. Joseph's Mission. From San Borgia he sailed farther south to Cape San Lucas in Baja California where he picked up another mare with foal. A book by the captain's son, H.W. Cleveland, points out that the skipper paid off with goods in exchange that would have commanded but $1.50 in Europe. So much for the price of horses in those days.

Captain Cleveland then sailed his vessel and live cargo to the Sandwich Islands, arriving at Kealakekua on June 21, 1803. There he learned from John Young, the *haole* (foreigner) advisor to King Kamehameha, that the monarch was presently over on the island of Maui. Disappointed at missing the king, Cleveland sailed northward along Hawaii's shore and put in at Kawaihae. There he delivered a mare with foal in June 23, 1803.

Captain Cleveland's report reads: "This was the first horse that ever trod the soil of Owyhee (Hawaii) and caused among the natives, incessant exclamations of astonishment."

Cleveland then transported the remaining animals over to Lahaina, Maui, where he left them with King Kamehameha and the latter's other advisor, Isaac Davis. Ironically, Captain Cleveland's report reads that the king displayed little or no appreciation for beasts. A thudding irony!

In fact, Captain Cleveland's description of Kamehameha and the event is best summed up in the captain's own words describing the king as: "Large, athletic but absent — he did not reciprocate our civilities." As for the other natives who observed this strange live cargo, Captain Cleveland wrote that they: "Expressed much wonder and admiration as was very natural on beholding for the first time, this noble animal."

It was a fantastically small beginning, it's true, but the animals were allowed to run wild and, in time, their progeny and other imports increased and multiplied in an awesome way. The horses were not large in stature, but they were of a rugged variety and well adapted to the kind of terrain where they were released. Then, through the years of freedom on land made up of towering mountains, deep canyons and lava-ribbed slopes, the animals scrabbled for existence and grew hardy, tough and agile. Particularly did the ones on the Big Isle of Hawaii thrive and multiply. Those wild mustangs became known as "Mauna Kea horses," and the herds were fast, elusive, and always scatting out of sight the instant a human being appeared on the horizon.

In that the first cattle were brought into the Islands in 1793, it gave them a head start of ten years over the first horses, for the latter were not introduced until 1803. And it took almost another thirty years before there were any vaqueros to help handle both expanding herds, for those

The First Horses

expert Spanish cowboys were not brought over from California until 1832.

From time to time throughout the years there were many further importations of horses from California and elsewhere. One documented shipment is found in 1828 when the French vessel *LeHeros* brought seventeen horses from San Diego, California. They sold for from $85 to $100 each in the Islands — a far shout from what they could begin to command on the mainland. California, then a colony of Spain, was surfeited with horses and, in that region, even a good horse could hardly bring the equivalent of $25.

Most of the early horses brought to Hawaii were mustangs — a corruption of the Spanish word *mestano*, meaning "wild." The so-called mustangs were originally the troop horses of the conquistadors. The animals were tough as quitch grass, deep-chested, long-winded, swift, yet heavy enough to hold half a ton of leaping longhorn on the end of a rawhide *riata*. They were also called *bronchos* — a name for unbroken early California horses, meaning "rough."

It is said that they were mustangs of a kind that were seldom curried, seldom made pets of, seldom coddled, Wild and wooly they were. Almost indestructible, they would take a roll, rise and shake themselves, fill up on *alfileria* — and immediately they were ready for anything. It is reported that they gave no quarter, and expected none — a thing which engendered the classic saying among the vaqueros: "Plenty of horses after I am dead!" The vaqueros simply looked out for their own skins first.

With hardy mustangs of that calibre, the vaqueros on the mainland resorted to Spanish bits and spurs that were a caution, to say the least. Ironically, those hard-riding Spanish-Mexicans who had to depend so heavily upon horses for their living, were inclined to feel that the smartest horse was at the same stage of mental development as the barnyard sparrow. They reined them accordingly.

Big, raw-toothed spurs were the order of the day, to say nothing of the heavy spade bit of conquistador design. Those two items of gear kept the meanest and contrariest of horses in line. No horse could argue long with that merciless spade bit, nor could the animal remain rigid or stationary when being raked by a biting long-spiked rowel. The vaquero felt that, without the advantage of the heavy restraining Spanish bit and the sharply-pointed rowel, only a compulsive gambler would bet on what a mustang would be apt to do. After all, with a string of half-broken cowhorses, a vaquero might get around to the same horse but once a week; maybe once and never again... So, the vaquero necessarily wanted to be totally in charge of his mount. He never knew when getting astride a horse, but that the animal might prove to be as personable as a panther.

In truth, it wasn't often that vaqueros had to utilize the brutal spur, for the very threat of the ever-present instrument was enough. The ugly rowel ultimately went into disuse as horses became better trained. The selfsame thing went for the Spanish spade bit. It was likely the best and most practical bit ever invented, but because misuse of it by a cruel rider could wreck a horse, the device was finally outlawed for all time. This was humane and this was good. Today's horsemanship is consequently on a higher, more merciful plane.

Those imported animals had roamed the American continent for 300 years and had come to be commonly called Indian ponies. Throughout the years of man's westward migration across North America, it was the mustang that carried pioneers, pony express riders and cavalrymen. The breed developed into the legendary cow pony of the Southwest. It stood to reason then that, transplanted to Hawaii and becoming the little Kanaka mustangs, they were sure-footed and agile. They stood between 12 and 14 hands high and weighed from 600 to 900 pounds. They had the capacity to carry heavy loads, climb rugged hills and work hard all the solid day. Today they are practically extinct.

A Dr. David Woo of Hilo, Hawaii, is currently striving to see that the Kanaka horse "runs again" in Hawaii. He points up the rarity of the Kanaka horse by saying, "The last wild Hawaiian horse was caught on Mauna Kea in 1924. He was a black stallion and his tail was so long it trailed on the ground. It took three cowboys to bring him down the mountain. He was bred to other strains and that was the end of purebred Kanaka horses."

In fact, for six years Dr. Woo has combed the Big Island areas of Kau, Waipio and Kona for near-purebred horses of the original strain which are now just about nonexistent. At the moment he has been successful in coming up with four stallions and three mares through his hunt-and-breed project. He takes delight in explaining: "Their conformation and small, hard hooves show the unmistakable imprint of the Barb-Andalusian influence of the Spanish horses which the conquistadores rode in their conquest of the Western Hemisphere around 1500."

The program he has outlined for himself is no simple task, pays no financial dividends and, fortunately, is something of a labor of love. The gentleman first took this avid interest in horses when practicing medicine at the Parker Ranch 30 years ago. Today he keeps his Kanaka horses on his Double U Double O Ranch at Ahualoa, and is prideful over his small nucleus for a someday herd of the fine, hardy little animals. He has them registered under the name of Royal Mauna Kea Mustangs.

Dr. Woo feels that it is still not too late to bring on a renaissance of the animal. He has also ransacked the islands of Kauai, Molokai and Maui for further remnants of the old wild horse herds. They're noticeable for their absence. In one lucky instance in Hawaii's Kohala District, he managed to trade two palomino horses for a Kanaka stallion.

"I found him working with a mule train, packing taro

out of Waipio Valley," he explained. Then at another time in Hawaii's Kau District to the south, he picked up one more much sought-after Kanaka horse. "One of the mares I bought," he said, "was being used as a saddle horse for tourists to ride scenic trails in the volcano area."

Yes, today Dr. Woo is a one-man equine crusade in Hawaii bent upon bringing about the restoration of that once-plentiful Moana Kea (or Kanaka) horse which roamed the Island terrain. Horsemen applaud him for his fine efforts.

But getting back to early history of the importation of horses into the Islands, it wasn't long before better blooded animals were wanted on the vast cattle ranches that were being established. On April 23, 1838 Captain Joseph O. Carter brought in a load of fine animals from California on his ship, the *Rasselas*. Many others were brought into the various Islands. The October 1936 issue of *The Friend* carried a report by Emil A. Berndt stating that all livestock in Hawaii in 1844 was on the big increase — stipulating that horses were part of the expansion. Some specifics are found in G.P. Judd's 1853 report to the Royal Hawaiian Agricultural Society. He pointed out: "Captain John Meek imported a grey mare in 1822," and "Reverend Ellis imported an English mare from Tahiti." These and many other importations from other regions were soon improving the quality of horses on the ranches where top mounts were mandatory.

As for the inferior horses from uncared-for stock which still ranged the land, some ironic aspects of the situation began to appear. It was brought to the attention of the Royal Hawaiian Agricultural Society by R. Moffitt that horses were becoming a positive nuisance. He described them as "useless horses which are increasing everyday and consuming ... the food which might be better employed."

In fact, the Island Legislature elected to pass a "Stallion Bill" in 1852 which imposed a tax on all horses. Its reasoning was that so many natives were counting horse ownership as a status symbol, and the attitude was self-defeating. For example, the Horse Committee of the Royal Hawaiian Agricultural Society claimed that "natives consider themselves rich as soon as they become horse owners, and will neither labor for themselves or for anyone else." Horses had really come into their own with the Hawaiians.

But it was something else among the wild horses that still galloped and grazed in the hinterlands. As early as 1851 a census of horses in the Islands showed 11,700 afoot. A great many of the animals that had been turned loose on the Big Island had congregated in the Mauna Kea region, and their disappearance from that high country wasn't a fact until the early 1930's. A story that touches upon this appeared in the *Pacific Commercial Advertiser* on February 1, 1904, which reads:

"Not long ago a herd of wild horses appeared near Waimea and old Tom Lindsey, followed by native cowboys, gave chase. They galloped after the wild animals up the steep side of the mountain. When the terrain became too rough, Tom Lindsey followed them alone and finally turned the herd and passed them back to the waiting cowboys, a twenty-five mile ride in all."

Just try to find a wild horse today in the Islands...

VI

UNIQUE COW COUNTRY

To give some conception of the kind of country in which Island cattle history was born, it might be best to portray Waimea on the Big Isle — the very cradle of the Hawaiian cattle industry. Quotes from old responsible archives tell us much. One of the best and most accurate is found in a September 10th, 1836 issue of the *Sandwich Island Gazette.* Here, in part, is some of it:

"The surface of the country is comparatively level, and generally rocky in greater or less proportion according to the age of the lava, of which the whole island is composed. In some parts the clinkers look as if they ejected but yesterday from the fire, black and bare, without a single blade of grass or a solitary lichen; while in other parts there is a slight soil of several feet deep consisting apparently of decomposed volcanic mud. The northeastern part is the most fertile, and the only part cultivated, the plains being for the most part covered with long rank grass, with but little wood.

"The northern mountains rise to the height of 5,000 feet or more, and are thickly wooded, abounding with large trees of various kinds, but the largest and most plentiful is the ohia lehua; here this species of Eugenia appears generally to take root on the top of the fern trees at a height of 20 or 30 feet from the ground, the roots shoot downward and reach the earth, the fern tree thus trampled on gradually decays and leaves the tree supported on a number of arching trunks with a vaulted space beneath, which, with a little trouble forms a good shelter for the benighted traveler, or wandering woodcutter.

"Much rain falls in these mountains, which, on this side, descends in three streams to the plains below; the first descends at the northeast extremity of Waimea near Puukapu, it is small, but seldom fails; it divides into two small rivulets, one running easterly toward Hamakua, is soon lost in the swampy woods, the other, westerly along the plain; the second is on the north part, and descends at Waikoloa, passes by the Mission House, through the village scattered about the neighborhood, and taking a southwest direction, passes by Lihue on to the sea, which however it seldom reaches, excepting after heavy rains. It then flows into the sea between Kawaihae and Puako. It is the largest stream, and waters the whole of the cultivated parts. The source of this stream is near the head of the valley of Waipio, whence it pursues its course through impassable woods, till it begins to descend to the plain, when its course becomes tortuous through a deep ravine, between two spurs of the mountain, their sides nearly perpendicular, being thickly covered with wood to the height of many hundred feet.

"The bed of the stream is broken and rocky, here forming a series of beautiful cascades, sweeping over and around stupendous rocks, there, turning a projecting point, resting in a deep still pool — tumultuous and turbulent it pursues its romantic course to the plain. This stream formerly abounded with waterfowl, especially ducks. The third and most westerly stream, descends to the plain at Keaali, by a majestic fall of near 100 feet, and although a considerable stream at first, is soon lost in the arid soil of this part of the plain, when after much rain, it reaches the sea; it is a little to the northward of the other.

"These three streams water the parts capable of cultivation at the foot of the hills, which are carefully preserved from the encroachments of the cattle with which the plain abounds, by a high strong stone wall many miles long.

"Just below the woods, cultivation commences; with little trouble the natives might always have abundance, but they prefer to trust almost entirely to the natural productions of the hills, such as wild plantains and bananas, wild turnip, sweet potatoes, raspberries, etc. *Mau* (Sadleria Cyatheoides), *KI* (Dracena Terminalis), and other roots and herbs are ate (sic) only in times of scarcity. There are some fields of upland taro on the sides of the hills, but the numerous acres of cleared unplanted ground that show the former extent of cultivation, tell a woeful tale of diminished population or of increased idleness."

Then with reference to Kawihae Bay, the port just west of Waimea from where the cattle have been shipped for lo these many years, the writer had this to say: "This place (Kawaihae), here in 1836, however barren, has its attractions, for the natives and foreigners; for the former it has its salt works and is the best place to purchase fish on the whole island. For the latter, it has its salubrious climate and tepid baths. But a few years since this was the headquarters of the chiefs, consequently the metropolis of the island; now, heaps of stones thrown into confusion by earthquakes, mark the sites of the houses.

"In this district you may find almost every variety of climate, from the oven-like dryness and heat of Kawaihae, to the cold wet rawness of Puukapu, and the freezing snows of Mauna Kea. But as there is but one inhabited spot inland, the observations here made must be supposed to refer to that. The thermometer, under the shelter of a house, ranges during the spring and summer months between 62 degrees and 75 degrees, but if exposed to the wind and rain, the evaporation will frequently cause it to fall as low as 47 degrees. This place, as a residence, is in many ways unpleasant; during a great part of the year it rains abundantly, with much wind, a nasty drizzling, soaking rain — a pure Scotch mist. During these times, which not unfrequently last for several weeks, exercise out of doors is extremely unpleasant and, to an invalid, impossible. At other times during the dry seasons, walking is equally disagreeable, for, from the lightness of the soil, you sink ankle-deep at every step. And if you are bold enough to face the wind, you get your eyes filled with dust, which the wind whirls along in clouds; cold and chilly, you seek repose and shelter in a house warmed by a charcoal fire.

"As the clouds roll along the vast expanse of the Pacific, they strike the bold coast of Hamakua, 2000 to 2500 feet perpendicular above the sea, they there discharge their contents. And as the country rises toward the center of the island, the wetter it becomes, but the instant the descent commences, the quantity of rain diminishes, so that at last the seashore on the western side which slopes gradually to the water's edge, is left dry and barren. The earth becomes very much heated and, being seldom cooled with rain, preserves an even temperature of about 80 degrees average the year throughout, with a degree of dryness almost incredible. The only water to be procured here is a brackish and very warm kind; it oozes out from beneath the rocks at low watermark along the whole coast.

"The population of Waimea is far from being numerous. Its numbers are constantly fluctuating, corresponding with the movements of Governor Adams Kuakini, who frequently resides here. Many foreigners have lately gone to reside there, principally attached to the cattle farms, or mechanics, whose services they require. Leather is made in quantities equal at least to the consumption on the Islands. Waimea harness and Waimea shoes are notorious for their strength and durability. The articles principally exported are livestock, salt, jerked beef, hides, tallow, leather, *mamaki kapas*, feathers, koa plant, etc."

VII

THE EARLY BULLOCK HUNTERS

By 1823 the "wild and ferocious herds" (as two historians term them) were being hunted for meat in a big way. Their old records tell us that those first animals which Vancouver had dropped off on the Big Island in 1793 had so increased in numbers and changed in temperament that they were a constant danger to people, crops and forests. The British sea captain had chosen this (as he worded it) "great tract of luxuriant natural pasture" so that the beasts might "roam unrestrained and increase and multiply far from the sight of strangers." Of course, the *kapu* which King Kamehameha the Great had put upon the animals had certainly carried out Captain Vancouver's plans many more decimals than either of them had anticipated.

In those early 1820's the manner of hunting and preparing beef was more than a little primitive. Salt was carried all the way from Kawaihae beach into the mountains and valleys where the cattle were killed, then it was applied immediately to the raw meat. Following this, the beef was put into barrels and lugged anywhere from ten to fifteen miles back to Kawaihae where it was put aboard the sailing vessels offshore. True, some of the beef was consumed locally, but Hawaiians, for the most part, still preferred their fish and pork.

As late as 1825 there were still no passable roads for horses in the cattle country. But by 1830 the Hawaiian Chief Kuakini (better known as Governor Adams) arranged for the building of a carriage road in from Kawaihae. This road, despite its roughness and crudity, was of great help in getting substantial quantities of beef down from the wilderness and plateau areas. Of course, at a later date, all this was made much simpler when cattle pens were built in the Waimea region and the animals were fenced in for domestication or further shipment. But, for a time, the wild herds were beginning to be thinned out due to the year-round open season on them.

History records that there was a drastic letup in cattle killing for several years beginning in 1840. It was the year when King Kamehameha III (Kauikeaouli) proclaimed the first Constitution of the Kingdom of Hawaii, and many things were happening at court. The tabu on cattle killing was renewed from 1840 to 1844, although we don't know how absolute it was. It resulted in many bullock hunters being unemployed, but it did give the wild herds a chance to propagate and thrive once more.

Due to the cutback in bullock killing during those several years, the herds did multiply rapidly and, in fact, by 1858 it was estimated that wild cattle on Mauna Kea numbered 10,000. This despite much illegal killing, some poaching, plus a heavy toll of young calves slaughtered by wild dog packs. With reference to the big voracious dog packs, records show that these roving marauders even ventured down onto ranchlands and slew right and left.

But his hiatus on bullock killing brought on a change which had not been foreseen. Historian Kenway of this era had recorded that the old-time bullock hunter was disappearing, and there seemed no one to take his place. Kenway wrote:

"It demands no common amount of nerve in the man and sagacity in the horse to face and fight these monstrous, unruly creatures. Great tact and practice are necessary.

The Early Bullock Hunters

The tales one hears of hairbreadth escapes, desperate adventure, and fatal accident which have rendered Mauna Kea famous might put tiger hunting to the blush and make the capture of wild elephants seem a small thing. Strange that such an exciting occupation should so effect its professional followers, but perhaps with one or two exceptions there cannot be found a more slothful and useless set of people than the (now idle) bullock catchers. Whether because of the introduction of the Spaniards or the result of the occupation, bullock catching of the old times had a mysterious effect upon its followers and spectators.

"The natives enjoy such sport amazingly, and as they cannot now touch the wild cattle, a great deal of unnecessary excitement is gotten up among the tame ones; and Beckley's Boys, who attend to the government herd are known by the clouds of dust that constantly envelop them. Waimea of an evening is a perfect cloud of dust. The soil is remarkably dry, and so extremely fine that water does not seem to wet it."

So much for that period of time when cattle were not being shot or captured. It does, however, give us something of an additional picture of country which, in ensuing years was to ultimately become one of the greatest and largest cattle-raising regions in all the United States.

VIII

BEEF AT A PREMIUM

The trading of beef, hides and tallow for foreign goods really began to come into its own in 1830. Back under King Kamehameha's orders, the island chiefs had been pressing the natives into more and more cutting of sandalwood trees for the use in barter. Sea captains from all over the world had been stopping at Hawaii's shores and exchanging merchandise for the fragrant wood; particularly had the Orient been using it for furniture making and incense burning. The sandalwood cargoes had also been traded for porcelain, tea, silks and a host of other items wanted by the Islanders.

Throughout the early 1800's the sandalwood exchange had worked fine, but there had to someday be an end to the sandalwood supply. The many years of harvesting practically ever sandalwood tree and sapling had about denuded the region of this long-time bonanza. By 1830 the Hawaiians were hard put to come up with anything as satisfactory as sandalwood for a bartering medium. However, they were delighted to learn that the foreign trading ships were happy to accept beef, hides and tallow as a substitute. Fortunately, the hills, valleys and slopes fairly swarmed with the wild cattle.

Meantime, the monarchy had changed twice; King Kamehameha the Great had died in 1819; his eldest son, Liholiho, reigned as sovereign for two years as Kamehameha II, and died in 1824; then in 1825, Prince Kauikeaouli, another son of Kamehameha I, had been proclaimed monarch. However, in that Kauikeaouli was but eleven years old at the time, Kaahumanu, the late King Kamehameha I's favorite wife, continued in the regency during the new king's minority. She, in short, proclaimed herself *Kuhina Nui,* or prime minister, and important decision making was naturally left to her.

Premier Kaahumanu saw the foolhardiness of the continued sandalwood harvesting, and put a tabu on the further cutting of it. She had seen where the chiefs had been forcing their people to cut the trees at the cost of neglecting their taro crops and their fishing. This had led to a serious shortage of food on the land, consequently an order went out putting a death penalty on any additional sandalwood cutting by the commoners.

So, beef became the chief means of barter and trade. It was also doubly fortunate that the wild cattle were available, because Hawaii's *alii* had gotten themselves into quite a financial bind. The lack of sandalwood had not restrained those of royal blood from continuing to buy foreign products. They had resorted to buying on credit. And now they were heavily in debt for goods like household fixtures, silk, wool, cotton goods, jewelry, guns, ammunition, et cetera. Merchandise such as this had become highly important to them, and they had gone somewhat berserk in all-out pruchasing. Obviously, they had thought their source of payment was as dependable as gravity. So, now they began leaning heavily upon their supply of cattle as a medium of exchange. For too long they had resisted change like a monastery.

Fortunately for them in this area, a new tremendous market for salt beef had come into being; the whaling fleets had begun their 50-year period of Island calls. The vessels

Beef At a Premium 49

wintered in the Islands and replenished their supplies of water, wood, fresh greens and beef. The *alii* went on buying as if there were no tomorrow. So, despite the seemingly bottomless needs of the many sailing ships, the indebtedness at the royal court continued to escalate. The Hawaiian semifeudal system had put the land through an economic wringer.

The upshot was that it became mandatory that the volume of beef be increased for bartering purposes, and this necessitated a search for improvement in the handling of cattle. There was, in truth, an efficiency and employment problem on the cattle ranges. One of the big steps in the direction of solving this cattle dilemma was the importing of a new breed of man skilled in horsemanship and expert in cattle work.

IX

THE ADVENT OF VAQUEROS

In 1832 King Kamehameha III sent a high chief to California for, among other purposes, to bring back Mexican, Indian and Spanish vaqueros to teach the Hawaiians the art of properly working the ranges. Of course, the long line of Spanish-Mexican vaqueros had originally been trained by the transplanted stockmen who had learned their husbandry upon the pastures of Spain. To be precise, the Spaniards taught the Mexicans, and now the latter would be brought to the Hawaiian Islands to teach the natives.

Three wiry, slat-thin cowboys of Spanish-Mexican descent were brought over to the Islands; by name they were Kossuth, Louzeida and Ramon. They started working on the Big Island out of Hanaipoe on the slopes of Mauna Kea, just twelve miles east of Waimea. There is no record extant of their exact reaction to finally being back at their trade after being so long at sea, but the hovering smell of crushed grass must've had them in a faint of memory. After all, the Waimea region is rich in vast reaches of lush grass. So, with the vaqueros, it must have been fine, just fine. Too, they were once again exposed to the listenable music of the wind on the mesas, and that must have had a touch of home in it. Even the far-rolling hills and travel folder mountain scenes must have been bittersweet as camp smoke to them. And recognizing the land as being prime cattle country, the whole region surely had become their *Pinta, Santa Maria* and *Nina* rolled into one. Even today the beauty of Hawaii makes many a *paniolo* wonder what he has done to deserve it all.

Those first imported vaqueros were something to see! And they probably would have been more than a little surprised to know that they would forever after be known as *paniolos*. To the Hawaiians, those strangely dressed *haoles* (foreigners) were something clear off this planet. Everything about them was baroque and exciting; everything from their brilliantly-colored woolen ponchos to their slashed leggings. Their brightly-hued sashes, jangling spurs, colorful head bandanas, broad-brimmed, floppy sombreros — all those accouterments were dramatic and tended to spellbind the natives.

The organizing and the training began. Things were a mish-mash to begin with. The vaqueros had to first select the better available horses and break them into good working animals. This was no sinecure, as pointed out, for the horses were direct descendants of the tough mustangs, or broncos, of the mainland's Southwest. They were difficult to bring to halter. In fact, the Hawaiians called them *Li'o* (pronounced lee-oh) — a Polynesian word signifying wild-eyed or wide-eyed. In time, *li'o* actually became the Hawaiian term for *horse*. Old-time Hawaiians still refer to them as such.

Immediately the vaqueros began imparting all the tricks of their trade to their new brothers. They were soon demonstrating horsemanship that was as efficient as it was spectacular. Their black magic with the lasso was astonishing. Their complete control of wild, cantankerous steers was a lesson in strategy and generalship. And the Hawaiians rapidly became avid students; being naturally athletic, robust and courageous, they took to cowpunching the way gamecocks take to fighting.

Although Hawaiians are normally leisurely and have a habit of wearing life like a loose garment, the ones who chose to become cowhands went at the profession like martyrs with a mission. Their zeal for mastering the unbroken horses, was such that often when trying to break one to saddle, they would tie themselves to the animal's back. The vaqueros considered this kind of dedication as suicidal. And sometimes when the beast deliberately or accidentally rolled over, the rider was killed or maimed. But such was the consecration of the Hawaiians trying to learn a new way of life. They meant to make the magnificent most of their opportunity.

The importance of the arrival of California vaqueros to assist in the capturing and handling of the vast hordes of cattle is particularly pointed up when we consider the incredible growth of the wild and domesticcated herds. By this time, their numbers bulged beyond belief. The increase was particularly remarkable in view of the slaughtering of them which had known no bounds for so long. As indicated, in 1841 Governor Adams had put a tabu on the slaughter of them just for their tallow and hides. An example of the massive killings that had taken place is found in the publication, *The Friend*, of July 1, 1843. It stated that 10,686 bullock hides had been exported at $2.00 apiece. Cattle raising and cattle killing had truly developed into big business in those early 1840's, and large operators like William French, John Parker, William Hughes and Janion and Green were the cattle typcoons of their time.

In retrospect, it's incredible that the herds of wild cattle continued to increase despite all the depredations that befell them. Not only were the cattle being slaughtered by the thousands for their beef, hides and tallow, but there was that additional wild dog factor that took a heavy toll of their numbers. Samuel S. Hill explained in his *Travels in the*

Sandwich and Society Islands: "I learned from friends here (Waimea, Hawaii) that it was believed that there were 100,000 head of wild cattle in these mountains which all had proceeded from one or more pair left by Vancouver on one of his memorable visits to the Islands . . . their number, however, was now in a fair way of being much reduced by the wild dogs which were very numerous and subsisted on the calves, few of which the cows were any longer able to raise. A quantity of poison had been imported by the Government for the destruction of these voracious intruders and a large portion of it had been forwarded to one of the native officials in this Island with instructions to distribute it disguised throughout all the temperate regions of the mountains."

Some additional conception of the cattle increase is found in *Thrum's Annual* of 1894 where a paragraph deals with the Islands in general in the 1840's and 1850's. It explains: "In some districts agriculture was entirely ruined by the encroachment of herds of cattle chiefly owned by foreigners. These herds were allowed to increase without limit until large tracts of country were completely overstocked, thousands of acres of fertile land laid waste and the rights of the native tenants literally trampled underfoot. The result was that the people in these districts became discouraged and gave up the contest. In 1851 fairly good cattle on Kaui were sold at $2.00 a head. Boiling works were erected in several places where cattle were tried out for their hides and tallow."

In short, had it not been for the arrival of the vaqueros to make cattlemen out of Hawaiians, it boggles the mind to think how much more out of hand the cattle situation might have gotten . . .

X

LASSOS

Thorough craftsmen that they were, the vaqueros took care in instructing how to braid the rawhide quirts and lariats (or lassos) — an art in itself. The lassos, particularly, had to be strong and functional. The rawhide rope (*kaula ili* in Hawaiian) of the *paniolo* was as important to him as a hammer is to a carpenter. His constructing of it was a labor of love, for there would be times when he would have to depend upon it for his very life. He put deep devotion into the making of it.

He demonstrated the making of the lasso by selecting a flawless, strong hide, the laboriously scraping off all the hair from it — the *koe i ka hulu*. Then, with a razor-sharp knife, he cut a spiral strip about a half inch wide from the very center of the hide to the outer edges. It took rare skill to cut the length true and even. Upon finishing the spiral strip, he showed how to divide it into four even strands; this involved much careful and dexterous cutting. For a very special fancy lasso, he sometimes even cut the original strip into eight strands — an adroit trick when one could do it. Then, using cattle fat, he oiled and softened the strips by hand, working it in until the lengths were flexible and pliable. The final operation consisted of braiding all the strips into one single rope, making it alive and sinewy so that it would snake out true to its target and hold like a length of steel cable. Next, he affixed a metal ring to one end of it, then fashioned a tight nubbin of a knot at the opposite end so that the whole length could not unravel. Some of these *kaula ilis* he made over a hundred feet long; others, to suit taste, a deal shorter. Much depended upon in what region of the country the lasso would be used: short-rope work in tightly confined areas, and long-rope work in open country.

The finished lasso was then coiled and attached to the *okuma*, or saddle horn, with a small length of looped leather, which was also fastened to the saddle tree. This short length of leather, called the *kaula hoopaa*, which held the lasso, was slipped through the coil several times and the loop was also placed over the saddle horn. However, in cases where a *paniolo* wished to be prepared for instant use of the rope, the coil of *kaula ili* was merely placed over the saddle horn and left there without binding. Today's modern *paniolo*, by the way, seldom works with a rawhide lasso. The present generation uses linen, hemp or one of the new synthetic fibres like acetate, dacron, etc.

But history shows how the instruction of those patient, skillful vaqueros had paid off. The Hawaiians were astute and earnest students and they learned well. They mastered the making of the lasso, then they demonstrated that they could be deft in employing it. The major obbligato in "rope" work is timing, and the Islanders soon proved their talent in this direction, too. If their race had fine timing for the world's greatest surfers, spear-throwers, swimmers, etc., surely they had the selfsame fine timing essential for roping. So, as the various races crossed and recrossed with their blood lines, the vaquero knowledge and skills were passed on and improved upon. A new breed of cowman developed, and he became all that the vaquero was — and more.

XI

SADDLES

Although the horse and the lasso were of prime importance to the *paniolo*, never could he discount the saddle. It was all part and parcel of the machine he operated with such precision. Of course saddles didn't come into use until long after the first arrivals of horses; twenty or thirty years, or thereabouts. The Hawaiian Historical Society's 1892 Annual Report gives the nod to the Mexicans for the introduction of their own saddles. It is more than a little difficult to put the finger on the precise date when saddles were first put upon the horses of Hawaii, but a piece in the publication, *The Friend* of January 4, 1850, does make reference to saddles. The article, "Visit of French Sloop-of-War *Bonite* in 1836," tells of a group from the ship going ashore at Kealakekua and obtaining horses and a guide at Kaawaloa. It explains: "Some of the horses sent for our use were furnished with English saddles and others with clumsy Mexican saddles."

The Mexican-Spaniard saddles built by those old-time saddle makers had no tacks or nails in them. They were hand-pegged, sewn and planed by hand, and each one was done to the individual's taste. Double sections of strong neleau wood were shaped and hand-planed into the base for a saddle tree. The two underboards were stretched over tightly with rawhide and hand-sewn with goat hide. A solid piece of wood served as the *okuma*, or pommel, and this had to be particularly stout to withstand the mighty strains and stresses to which it would be subjected. The *okuma* was then tightly covered with rawhide. The neck of it was sheathed with additional layers so that it would withstand the friction and pull of a tightly strained lasso.

The *okuma* was then temporarily tacked to the underboards so that holes could be drilled through the two jutting ends and pegged with neleau wood about a half inch thick and four inches long. After the wooden pegs were knocked tightly into place, the temporary tacks were pulled out, and the *okuma* pegged firmly to the underboards.

The backboard was the next installation — a hand-planed arced section of wood covered with rawhide and hand-sewn with goat hide. Just as soon as this section was temporarily tacked to the two underboards, holes were drilled and the pieces hand-pegged into place. Then, with the underboards, the *okuma* and the backboard brought together into one single unit, the leather parts were added. This consolidated the underpiece of leather, the back piece, the stirrups and the other essential parts.

Depending upon individual taste, sometimes a large sheet of leather (the *lala*) was spread over the saddle simply for decorative purposes. It wasn't functional, but did furnish additional seat padding. Taste, too, was often displayed in the *paniolos* affixing their own leather markings and patterns which they hand-stamped and hand-carved to supply the gear with a personal touch and give it individuality. As for the saddles built for "shipping" — that is, where the mounted horses were compelled to swim into the surf to work cattle up to the offshore boats — they were constructed of a bare amount of leather because cowhide disintegrates so rapidly after exposure to salt water.

All these intricacies and complexities of a cowman's trade were carefully taught to the Hawaiians by the vaqueros. What the natives didn't learn through word of mouth, they learned by observing; and they certainly proved to be eager, dedicated pupils. They rapidly became horsemen and cattlemen in their own right. Physically, they had all the equipment, and as for courage — not a one of them would hesitate to go up against a mountain slide.

XII

HORSES, TERRAIN AND HORSEMANSHIP

It was the horsemanship of the vaqueros that engendered the most excitement on the part of the Hawaiians. The Spanish-Mexicans were past masters in the saddle. But horsemanship takes two — a man and a horse; and the horses had to be caught, broken and trained. The vaquero displayed the patience of a padre and the skill of a scientist. As pointed out, those wild horses dated back to 1803 when Captain Cleveland first brought them over and turned them loose in the Islands. Running wild in wild country for so long, they were wiry, strong and freighted with stamina. As to breaking to saddle and rein, they fought like demons against man making beasts of burden of them. They became named "Mauna Kea horses," and it took strong hands to master them. Broncoes they were with a big bag of tricks, and a bang in every bag.

The vaqueros were soon demonstrating a form of horsemanship to the Islanders that was superb. Rider and horse had to work together like one unit — a flesh-and-blood machine that functioned with precision and efficiency. But capturing the better mustangs in the thickets and forests of the island offered the biggest problem. Breaking and training them was secondary, yet that was hard enough. The progress and prosperity of the ranches depended upon good horseflesh — and lots of it. The fact was that efficient, first-class cow ponies were at a premium.

Another thing much to the credit of the riders was the manner in which they coped with the rough, treacherous terrain of Hawaii. Anything less than excellent horsemanship could result in pure disaster. Lava-strewn ranges offered murderous iron-hard surfaces to ride on. And horses' hooves faired badly. *A'a* — lava in its rough, crumbling stage — is a frightful obstacle to negotiate. Molten lava has left hugh pockmarked and crevice-ridden regions that make for thousands of leg-breaking traps. It was vicious terrain then — as now — and a fearful challenge for any horse, trained or untrained. The spiky tangles of *kiawe* (pronounced kee-av-ee) offered additional complications for rider and horse, and made cutting out or lassoing cattle a big-time trouble. In fact, because of the uncommon roughness of the terrain and the belligerence of the cattle, it took a sure-footed horse to qualify for training. It was mandatory that the animal had to be fast, strong and smart.

The vaqueros were already steeped in rare talent, but the horses had to be trained to match their riders' skill. Then to transfer this ability to the Islanders required a patience and a teaching capacity that was unique; and, bless them, the vaqueros had it. Those knowledgeable vaqueros had no trained quarter horse type animals such as were developed in later years, but they tried to pick and train the nearest thing to it. The selected horse had to be endowed with tremendous alertness, yet without extreme nervousness and "spooking" tendencies. The animal had to be powerful with fast reflexes — and certainly not too big to be maneuverable in tight situations. On top of all this, it had to be strong and agile under the saddle, and capable of outrunning the fastest steers, whether the latter would be charging or escaping.

This was a lot to look for in a horse, but the vaqueros

were fussy about their mounts. They soon had the Hawaiians and the *hapa-haoles* (half foreigners) looking for the same thing in their steeds. It took a deal of serious searching, training and adjustment.

However, one thing was patent in those early days; and that was that, with the accelerating demand for more and more beef, the demand for cow ponies would continue to be pretty exhausting. It meant that too often the new cowboy trainees were riding what were called *pupule* (crazy) horses or *hapa-laka* (half broken) horses. *Laka* (fully broken) horses were at a tremendous premium. And again it demonstrates the dedication and courage of those early *paniolos;* they had to make do with what was available.

The better trained the horse, the better the results on the range. Particularly in tracking down the wild steers (*pipi* — pronounced pee-pee), so much depended upon the sensitivity and intelligence of the horse. In forested or heavily-brushed regions, the horse was counted upon to smell, hear or sense the position of the hidden *pipi*. And usually the horse's eyes and ears would come alert and point out the exact area where the cattle had concealed themselves.

Instantly, the *paniolo* would dismount and cinch up the saddle girth (*kaula opu*), remount and spur in to work his coiled lasso on stampeding animals. His selection would be a bull, for it would represent more beef on the hoof, plus the fact that the hide would be bigger and better. So, the mad chase would commence and the nimbleness and intelligence of the horse counted for so much. Or, as was often the case, a bull would swing into action and charge horse and rider with lowered horns. This called for fast, desperate maneuvering on the part of both the *paniolo* and his mount. Yes, the two of them had to be a team all the way all the time.

In the event of no charge, the *paniolo* would cut out the steer he wanted, drive it into a clear area where he could do his rope work and bring the animal to a halt. Trying to wield a lasso in a labyrinth of mamane, chia and koa trees would often end in futility, so open, unobstructed areas were a must. Such a clearance, or open spot, is called a *kipuka* or *pa hu'a*.

At that point the *paniolo* would have to make an accurate, legerdemain throw of the loop before the steer could thunder across the clearing and disappear into the tangles on the opposite side. Good roping was of the essence, or the steer was lost. Providing that the throw was perfect, the *paniolo* would instantly take several turns of the line on the horn of the saddle to cinch it. His trained, co-ordinated horse would brake to a stiff-legged halt so that the lasso would come iron-tight. The fighting steer would strain at the long leash while the horse continued to back off to keep the strain on the lasso. The whole process was efficient and artful, and could be performed only by a trained horse and expert rider. The two of them had to be sychronized and machine-perfect, else they might be reduced to a bloody shambles. The steer could easily wheel and charge forward with bone-crushing impact, impaling either man or mount on lancelike horns.

Whether or not the steer charged, there would always be that berserk twisting and thrashing for escape, and the animal would be — as the Hawaiians call it — *eli ka lepo* (mad like hell). The *paniolo's* next move would be to work the steer over to a strong tree or stump to cinch the lasso around its base, and pull the beast flush to it. This would invariably bring on the moment of truth, for the *paniolo's* following move would be to quickly dismount and secure the steer's horns close to the trunk with a hand rope. This snubbing method sufficed to bring the steer to a standstill.

Animal after animal was caught and secured in this manner, and it made for arduous, dangerous work. A solid day of catching and securing wild cattle this way was man-killing labor. At the day's end, the *paniolos* returned from the thickets and made their tired ways back to headquarters. Horse and rider had had it for the day. Upon the morrow they would return with tame bullocks and tie them to the still-secured wild ones. In this fashion, with one domesticated bullock (called a "pin bullock") firmly lashed to each of the newly captured ones, the whole herd was driven back to the holding pens on the lower slopes. And there, some of them were made available for market at once, whereas many were selected to be added to the increasing domestic herd.

Sometimes the *paniolos* resorted to another mode of catching the wild cattle, and that was to wait for moonlight nights when the animals would come down from the higher regions to drink at the ponds. One small body of water on the Big Island at Waimea, known as Makini, was a favorite watering place for the wild ones, and there the *paniolos* made some of their easier catches; at least they didn't have to buck the thorny tangles and long, hard rides.

Another method which sometimes proved successful was to herd numerous domesticated cattle into and around wild ones until the latter were somewhat placated, then the *paniolos* would gradually work the entire herd, — wild and tame — back down to the pens. However, this didn't always work, for too often the wild ones were too wily for this ruse. The hard way was usually the surer way.

A thumbnail sketch of Hawaii's commerce in cattle is found in the *Sandwich Island Gazette* of Sept. 17, 1836. It portrays an even different manner of roundup from the one we've just described. It reads: "Pens for catching the bullocks on the southward side of the plain of Waimea... are built of strong posts of hardwood with crossbars of the same, strongly lashed together with strips of rawhide; they are all varying dimensions. From the entrance of the pen are two diverging fences built of the same materials, extending from a quarter to half mile in length; through this funnel a herd of wild cattle are driven by a number of horsemen and propelled on until they enter the pen, the entrance of which is immediately closed... The bulls are mainly killed for their hides and tallow, the flesh is generally wasted; part of the cows are slaughtered but their flesh is either eaten or jerked for the Oahu market or packed in barrels for ship use; the rest of the cows and calves are sent by ship loads to Honolulu."

Despite the new and strange conditions to which the vaqueros had to adjust, and even with the language barrier compounding their problem, those Spanish-Mexican centaurs managed to fit into the Hawaiian picture and convey their know-how to a race of people who had actually never seen cattle until 1793 — just thirty-eight years before the vaqueros arrived.

Also despite the vaqueros having to work with horses so inferior to the ones they had known on the mainland, they still managed to do their chores and teach the Hawaiians how to become the best of wranglers. Of course, today the *paniolos* ride mounts which are a far shout in quality from the wild mustangs of yesteryear. Hawaii now has the best. No finer horses are bred elsewhere. For years purebred stallions and thoroughbred mares have been brought into Hawaii, and many of the ranches have come up with riding animals second to none. Great bloods like Belgian stallions have been of incredible help. It is no uncommon sight to witness a *paniolo* astride a horse that is the equal of the finest bred polo ponies.

True, today where wheeled vehicles are practical, they use them, but working with cattle is still basically a horseman's job. So, on the winds is still heard the muffled *clop-clop* of horses mixed in with the sodden shuffle and snuffle of cattle. The creak of saddle leather mingles with the *slap-slap* of reins, and everything is movement and action. The smell of sweat is pungent and robust. And the *paniolos* with their call of, "Ha, pipi! Ha, pipi!" rises above the other sounds and noises. It's all a sort of music to those who listen. And it is a good music. Day or night it is good. There is a sort of poetry in it. At night the animals are ghostly objects moving along slowly. In daytime they take on a sharp reality, and one's mind goes back to the yesteryears when those Spanish-Mexicans were first imported from the West Coast to train and help the Hawaiian hands. To the credit of those first *paniolos* who left their homeland and came to these strange, distant Islands, they had to have towering faith and courage. Possibly they felt like the 15th century explorers felt when they embarked upon the unknown seas guided only by incomplete charts often offered as sailing directions: "Here be sea serpents and monsters."

Courage is where you find it; and the *paniolos* had it.

XIII

HAPPY INTEGRATION

The Hawaiians and the *hapa*-Hawaiians (part Hawaiians) became more than a little captivated by the personalities of these Spanish-Mexican vaqueros who had come to their Island to teach a trade. The expertise, daring and general character of these lean, hard-riding foreigners from California took hold of the Islanders' imaginations, and they soon became prideful in emulating the dress, horsemanship, ropework and manners of the newcomers. To the Islanders, these strange *haoles* (foreigners) were explosive with color. In a sense, the vaqueros brought a new world to Hawaii, and the Islanders took to them like bees take to blossoms.

The late Emma Lyons Doyle's fine documentary, *Makua Laiana*, affords us a colorful picture of the *paniolos* in their early days in Waimea as described by an eyewitness. It's an evening scene in Waimea's lone store, a thatched structure where the men often congregated. It reads:

"A bright fire was blazing in a cavity in the earthen floor, displaying in strong light the dark features of natives gathered around it in their grotesque attitudes. A group of fine looking men were leaning against the counter. They were all attired in the poncho, an oblong blanket of brilliant colors having a hole in the middle through which is thrust the wearer's head. The pantaloons are open from the knee downward on the outside, with a row of dashing gilt buttons down the seam. A pair of boots armed with prodigiously long spurs completed their costume. They are bullock hunters just returned from an expedition of ten days."

The early *paniolo's* way of living was necessarily simple to the point of starkness. Today his life has many refinements. The times are easier for him here in the '70's, thanks to newer and more efficient methods of cattle ranching. A lot of ingenuity has gone into the development of the ranches as we find them now, and much of it has been to the advantage of the *paniolo*. He himself has brought on many of the innovations, methods and devices which have eased his life. He has been living a meaningful life, and he has well earned his present easier life. In short, he has paid his dues.

He married in with the Islanders, bred in with them, and became part of them. Under vaquero training and association, a new Hawaiian cowboy has been developed, and the talent of the man has given backbone to the zooming cattle industry. In truth, the *paniolo* built the cradle of Hawaii's cattle kingdom. True, a certain few of them have battled their way to the top, then bottled their way to the bottom. These are the exceptions.

Changes have been rampant. Even the manner of dress is something else. Today the *paniolos*, except on festive occasions, are apt to wear anything in the way of clothes when plying their trade. There are conglomerate costumes of flamboyant aloha shirts, leather jacket, blue denim, or pedestrian sweat shirts. Dungarees and boots predominate. Kerchiefs generally hang loose except when the dust smokes up from trampling hooves, or swift winds blow sand and gravel like so much buckshot. Often the straw or felt hat is adorned with a lei of shells or fresh flowers; pansies are a favorite. The saddles are high-horned, and generally reminiscent of the old Spanish type with leather work done to suit the rider's individual taste.

And the names are polyglot, strange and thought-provoking; Sakades, Purdy, Lewi, Kimura, Quintal, Lindsey, Pelekane, Horio, Cumlat, Dowsett, Hui, Bell, Paro, de Silva, Ah Fong, L'Orange, Yamasake, Perreira, Christensen, Kihe. You name them, they're in Hawaii.

These are the new generation *paniolos* who do the herding, the branding, the castrating, the inoculating, the dehorning, the men who take their positions at the chutes and the chute gates to divide the calves from the cows and steers. They bring the magic to the roping, the fast, hard riding and the final shipping. And to prove whether or not they like their trade, with most of them, a vacation is generally just a rumor. Still, they stick. Many are in the winter of their lives — and still they ride! They talk proudly of the old days and of the newer days, and they create a towering tapestry of an eternal Hawaii.

XIV

A NEW BREED OF MAN

As heretofore described, purebred stallions and thoroughbred mares were brought in from the mainland to improve the *paniolo* mounts. That goes likewise for the cattle; the finest breeds of bulls and cows were imported to cross into the herds that ranged the hills and valleys. Particularly on the Big Island did this prevail, for that is where cattle ranching on a vast scale first began. The Hereford cattle were recognized the world around as the best of all breeds capable of foraging for themselves on open ranges and producing top grade beef. So there was the introduction of carefully selected Herefords to enrich the class of cattle on the land. Pure-blood Herefords were imported from such places as Wisconsin, Indiana, Missouri and Kentucky.

The change for the better was slow but spectacular. After a handful of years, a far better grade of beef was being raised and marketed and far better prices were commanded. Too, other ranch growth was being expanded through the development of carefully bred mules and draft horses. Whole paddocks were given over to this expansion. It meant that, in time, the *paniolos* were astride horses that would take top prizes in the best of horse shows anywhere in the world.

The same change prevailed in the breed of people; through intermarriage, association and general integration, a type of *paniolo* developed that became classic in its own right. The *haole*-Hawaiian union created a very special man. The vaqueros brought much to the Islanders, but the latter also had much to give to the newcomers. Where the Spanish-Mexicans were imparting valuable knowledge about the handling of cattle and horses, the Islanders were gradually educating their hot-blooded new friends in a Hawaiian tolerance and understanding of human frailities. No one could be around Hawaiians long before the aloha spirit would show its spiritual strength and begin to take over. *Haoles* from any country were better for it after experiencing considerable association with the Islanders; call it osmosis or what you will, but the effect has been notoriously healthy and gainful to character. As for the Hawaiian, he gained added drive in his heart without having sacrificed any of his native charm.

This new breed of man loved his work and made a religion of self-reliance. Top horsemen from far lands who visited Hawaii, marveled at what they saw in the Hawaiian *paniolo*. In the saddle, the latter was superb; his hard, graceful riding, his finesse with the rope, his technique with cutting out or herding cattle, his complete mastery of horsemanship in its every aspect — all these things marked him as outstanding.

Some of the finest and most legendary *paniolos* stemmed from the Purdy people. The bellwether of those fabulous families would be Jack Purdy. He had been stranded in Hawaii in the early days of King Kamehameha III. He and John Parker, the monarch's chief cattle killer, were soon in partnership trying to furnish the king with the beef so badly needed for barter with visiting foreign ships.

A New Breed of Man

In truth, by this time the wild cattle were in such numbers that it appeared they could never be measurably diminished. The demand for hides, tallow and beef as barter with the visiting traders had swollen to a point where it seemed to be inexhaustible.

Jack Purdy, like so many other *haoles*, married a Hawaiian woman. His wife, Hoka, gave him strong, beautiful sons and daughters who perpetuated the courage and skills of their parents on into these later years where the name "Purdy" became something to reckon with in the Islands. Some light can be thrown upon the kind of blood with which these Purdy people were endowed by taking the liberty of quoting an anecdote from de Varigny's fine *Fourteen Years in the Sandwich Islands*. It reads:

"We decided to ascend Mauna Kea. Everyone advised us to get Jack Purdy to be our guide and to supply us with horses. Jack Purdy is a type of a curious set of men in the Hawaiian Islands. He camps, rather than lives, in a large house in the midst of a plain. There is not a tree nor a flower in his enclosure which is incessantly trampled by twenty horses. He is an Englishman, forty-five years of age, and he arrived in Hawaii in 1834. At the age of ten he sailed from Liverpool as a cabin boy, and as a sailor of twenty he was shipwrecked on the coast of Hawaii, and there he remained.

"Jack is the best horseman on the Island, the most intrepid hunter of wild cattle, the man who knows the paths of the forest, and the passes of the mountains. He is an indefatigable walker and covers long distances without giving up. He is always sure to find food and lodging in the woods with his gun and hatchet. Jack would be perfect if he did not get very tipsy when he has nothing to do. That is his manner of resting. He is usually accompanied by four great dogs whose numerous scars attest to their bellicose disposition."

Jack Purdy's heroics are legion and by the time he died

and was buried at his home, *Po'o Kanaka* — "Man's Head" — on the Big Isle, he had fathered some sons that became *paniolos* second to none.

A prime example of to what heights the Purdys and other *paniolos* reached as horsemen and cowmen is well brought out in the way they acquitted themselves in rodeos. Probably the greatest of them all was one of the later Purdy boys, Ikua. He was a strong man, a tremendously well-balanced man, with a baroque sense of humor and a vast zest for life. His rodeo feats are still memoried things among old-time cattlemen.

Ikua Purdy was born in the shadow of Mauna Kea on Hawaii on Christmas Eve, 1874, and he proved to be a cowman of classic quality. Lean and wiry with spring steel for muscles, he carried himself like a man who knew where he was going. He always accepted life with enormous appetite. He early learned all the fine points of horsemanship, roping, tying, horse breaking and general cowmanship. While still only a boy, he worked the ranges and corrals with the best of the men. The way he rocketed his lasso out to fast-pounding steers was a thing to see, and much was expected of him in his maturity. He lived up to expectations — and more. In later years the other *paniolos* used to listen to Ikua as if they were in church.

His first big exposure to the general public was at the big Wild West Show of October 21, 1905 on a sunwashed afternoon in Honolulu. He was fantastic in every event in which he entered, and was particularly outstanding in roping. He set some kind of a record in roping, throwing and tying a wild steer in 58 ¾ seconds, and the throngs threw away their brains in applauding him.

On December 14, 1907, Purdy again demonstrated his rare skill at Honolulu by beating the world's champion rodeo man, Angus MacPhee. The latter's best time for roping, throwing and tying a steer was one minute, 10 seconds. The crowds again took the place apart in their worship of Purdy and his craftsmanship. Several other

Island *paniolos* also exhibited fabulous rodeo talent in the contests — and outdid the mainland cowboys by quite a margin.

Another one of the better artists with rope, horse and steer was Eben "Rawhide Ben" Low, also of the Big Isle. The tough, angular, mustachioed man had suffered the loss of his left hand while roping a wild bull in the Waimea region. It so happened that his wrist had become entangled with the rope and had so ripped the hand that it had to be amputated. Yet, at this later date, Eben Low managed to excel at roping despite his handicap of the one missing hand.

By this time, Eben Low, who was something of a leader among the *paniolos*, was so confident of the men he worked with that he wanted to have them demonstrate their unusual wares on the mainland. Pride was in it for Low as well as all the *paniolos*. In 1908 he arranged for several of them to enter the Frontier Day events at Cheyenne, Wyoming — the Madison Square Garden of those days for rodeos. The group was made up of Eben's brother, John, his half brother, Archie Kaaua, and his cousin, Ikua Purdy.

It was on August 22nd of that year that the Hawaiian contingent of *paniolos* showed Wyoming and the world what fantastic rodeo men they were. Those Island cowmen did things with horse, rope and cattle that swamped the mind. To begin with, the *paniolos* were dressed in their best vaquero style with a few Hawaiian touches added like flowers in their hat bands, and their aloha spirit was positively contagious. When they went through their roping and riding routines, they reeked with color and class.

Under a molten sun in a blue and cloudless sky, the colorful *paniolos* proved to be the sensation of the meet. Everything was hubbub and huggermugger as the stands exploded with sound at sight of the *paniolos*. Their dress, their talent and their personalities had captivated the crowd as well as the interest of the opposing punchers.

Ikua Purdy sat ready to tackle his first fleeing mainland steer. He was wearing the Mexican leggings introduced to Hawaii by the *Espanols* so many years back; his tight Gaucho-like pants and his floppy sombrero-type hat made him look like something out of a picture book. But the throngs hadn't seen anything yet. Likewise for the mainland wranglers. They had no knowledge that Purdy was in the habit of performing his feats at home under the toughest of conditions, what with thorny brush, rock, tree trunks, and scattered lava chunks on every hand. Treacherous terrain for sure. But there in Wyoming he had flat, unobstructed ground to work with, no fallen logs, no spiny *kiawe* bush, no fissures in the ground. He was in a position to perform mind-spinning feats.

There was the sudden release of the rangy young steer, the swelling of the crowd noise, the forward lurch of Ikua and his horse. Speed and timing were of the essence now. With his horse drumming the ground at breakneck speed, Ikua snaked his looped lariat out and snared the horns of the young steer. His well-trained mount hunched down to a grinding halt, went stiff-legged and brought the steer to a somersaulting stop. Purdy rocketed out of his saddle like a phantom and was at the struggling steer before it had regained its feet. All in one rhythmic, blurring motion he bound the animals's hooves with his short length of rope and flung his hands upward signifying, "Done!" The stunned rodeo aficianados blasted with applause, recognizing that this was something very special.

"Fifty-six seconds flat!" came the report through the pandemonium. The shouts and handclapping rose to a new crescendo. A craziness was on the crowd, and noise mantled it. Purdy bowed in thanks, adjusted his hat, and walked away with a knowing smile on his leathery face.

Ikua Purdy had captured the World's Championship steer-roping contest, and the fans had witnessed the making of some important rodeo history. What made Purdy's feat particularly outstanding was the fact that he performed it aboard an unfamiliar horse; the mount had been supplied by the mainland's rodeo association.

Archie Kaaua, too, scintillated for Hawaii by coming in second in time to Purdy. Jack Low placed sixth. All this competition had been against the cream of United States cowpokes, and it brought more than a little repute to Hawaii. Most people on the mainland hadn't even known there were any cowboys or cattle herds on those distant Islands in the Pacific. It was, in short, a great day for the Islanders, and they brought their trophies back home with justifiable pride.

But it stood to reason that those Island cowboys should do well in competition; after all, they had had the advantage of early *paniolo* training from those first imported Spanish-Mexican vaqueros. At least the training had been passed on down the line from grandfather to father and from father to son. It was their heritage and they had not defaulted.

The Purdys, like the Lindseys, represented a long line of top cowmen and horsemen. Islanders are still talking about another Purdy, one Jack Nae'a Purdy, son of the original Jack Purdy, who was a pioneer cattleman in the old days of cattle hunting before the advent of the Spanish-Mexican vaqueros in Hawaii. The old-time residents of Kiholo, near Kawaihae, still sing the praises of Jack Nae'a Purdy, and they say there never has been his equal. For years they've talked of how he used to swim his horse out into the sea when steamers stood offshore to take cattle aboard. Rough, pounding waters daunted him not when working cattle out to the boats for further transfer to the steamer vessels. Occasionally cattle broke from their lashings and panicked seaward. It took this particular Purdy to spur his horse into shark water to overtake the fleeing animal. In every department of cowpoking, he was a titan.

But the Purdys had company; right up and down the pike they had company. Good company. There were the

Lindseys — first-rate men, all — and a legion of other outstanding individuals like David "Hogan" Kauwe, Willie Kaniho and Harry Kawai. Over the years, men of their stamp and calibre have been riding range on Hawaiian soil and crying out to their cattle, "Hup, thaaaa! Hup, thaaaa!" Or when in dry periods on the dusty, open spaces, the yell has been, "*Ai lepo! Ai lepo!*" — meaning "Eat dirt! Eat dirt!"

To this day, all the described complications — except the cattle-swimming gimmick — still prevail on the larger cattle ranches, and good riding, good horses, good roping, and courageous men continue to be a must. Particularly on the Big Island, the heavy, suffocating dust stemming from moving herds is still a threat to *paniolos*. Much of the ranch weather on the higher slopes of Mauna Kea is crisp and misty in the nights and early mornings; the endless acres of cattle droppings become damp, then the sun blooms later and prairie land dries out. When a big, driving wind comes up, or herds of cattle are on the move, the droppings become mixed with dust and form a stinging element that blows into the eyes and nostrils of man and beast. It may sound minor, but the smarting of the eyes, nasal passages and throat can be insufferable. But the *paniolos* take it in stride. It is part and parcel of their lives.

In levity it has often been said: "If you want to see the cattle country weather, just look at a *paniolo's* face." It will truly tell the whole story. Too, it is often said that you can recognize a man from that vast high Waimea region on Hawaii or those high Haleakala slopes of Maui because of his pink cheeks — a condition due to the crisp weather. But probably much of it has to do with his happy state of mind and all-around good health...

The *paniolo's* life today is a good deal easier and safer than in times gone by, yet if you ask any one of the old-timers about his lot in the yesteryears, he'll quickly tell you that, though it was a hard life, he would willingly live it all over again. And he'll be the first one to confess that he, at least, always had the pleasure of working in a land of mind-stopping beauty. And that was a big plus.

In short, this has been the new breed of men born to the Island ranges. Good men, capable men, dedicated men — all of them helping to build a cattle kingdom in the Islands. They have built it, and with their progeny sweating at it today, it could be an everlasting good thing.

XV

THE *PANIOLO'S* LOT

The *paniolo's* lot was, and is, a hard, robust, healthy and exciting life with cattle hunting, herding, branding, shipping and marketing. It is many things in many ways. He meets all kinds of rugged weather — flashfloods, big, driving winds, thunderous rainfalls, and sometimes blasting hail storms that sweep down from the high-piled mountains and hills. Yet there are the better days — the typical Hawaiian days throughout most of the year when the weather is pluperfect and each hour in the open is pure holiday. He looks at the beauty of the country and it almost buckles his knees. At such times, weatherwise, a *paniolo's* lot is a sweet, sweet thing, and he goes at his work with missionary fervor. And the nights; they're starshot and serene with seldom a letup.

The enormity of some of the Island ranches comes as awesome in the eyes of most mainlanders. The *paniolos* take pride in the hugeness of some of the spreads. On Hawaii, when the thunder claps up there in the heights of Mauna Kea, it's a sound that makes the spine go stiff. The *paniolo* grins; it is *his* mountain. On Maui, when the nets of distant rain trawl slowly through the light and drain on the vast slopes of Haleakala, the *paniolo* shrugs his shoulders, pretending a casualness he doesn't really feel. Those are *his* slopes. All of the hauntingly beautiful country seems to be his alone. He looks across and beyond, up to the foot of the misty hills that fold peacefully into each other — and he feels as rich as a rajah. He knows that there are special moments in life, and this is one of them.

Few *paniolos* think in terms of ever swapping their trade for anything else. They're comfortable and at ease with their work, and somehow they even seem to love the daily calculated risks they are compelled to take. They don't even object when working the lowlands, and the cattle are moving torpidly in the heat and the dust. Something holds them there; maybe it's just that unnameable something which flows like a running stream from the men who work on the land, back to the land from which they sprang.

Every Islander who turned to cattle work was soon to learn that the career of a *paniolo* was a tough, demanding lot. His day started at sunup and seldom ended before sundown. Pay was small, so it took a dedicated man to hack it. He was almost constantly in the saddle, and the work with militant cattle was fraught with danger with little letup. At the end of each day's hard labor in canyon, forest or on the mesa, he had a slew of odd, but important, chores to take care of. He had to tend to his horse, look to the saddle and tack so that the gear would be ready for tomorrow's stress and strain — and this often involved major mending. He had to carefully examine his rawhide rope and make sure that friction and fraying had not weakened it.

Breaking and training a spare horse often occupied his "spare" hours. If he was without a wife, and on his own domestically, there was the thing of cooking meals and feeding himself. This is the measure of the *paniolo's* extracurricular duties; manifold and endless. It was a raw, hard life, despite the fact that most of the time the *paniolo* was working in a paradise of a land with unbelievably good weather. Much of this still holds true today.

Yet, with all the heavy labor and long hours, the *paniolo* is repaid with compensations that are as rewarding as dollars. Extra dividends come his way each day and they cannot be measured in cash. Nature herself in Hawaii pays off manyfold. A daily course of rare dishes — spiritual dishes — are the lot of the *paniolo,* and he relishes them with all his heart. The early morning sun on a prairie of high grass reaching out like a silken cloud is an apparition he never tires of. A great herd of Herefords on the move, flowing in a rich red stream, hypoes him up with its own brand of exciting stimulus. Idly he hears the *sish-sish-sish* sound of the animals's unseen hooves pushing through the knee-high grass. Tonic to the *paniolo.*

The cowhands have their last warming cup of coffee for the morning, the solid breakfast still puts the strength in their legs and arms — and they feel indestructible. No wonder that the sights this day are heady again. There's the smell of the saddle leather, the squeak of the gear, the snuffing of the horses as they clear their nostrils for more of the morning's crisp air.

Nearly every early morning the *paniolo* studies a procelain-pretty sunrise. Later he sees clouds building up, then breaking away and thinning. Later still, and looking to the heights, he sees what the winds are doing to the now ragged clouds as they scud up the slopes. And beyond the mountains are the high-piled white thunderheads like so many monstrous lumps of cotton. A mere turn of the head, and in the opposite direction is the trackless sea sparkling delicately like silver filigree. He might even watch lines of *nene* geese forming their flight tracery in the sky. He rides on, and he just might not return until the stars are deep in the black velvet of night. But it's a *paniolo's* world, his workshop — and he'd trade it for no other.

In passing, it might be interesting to point out that

practically all the ranches throughout the Islands are basically Hawaiian institutions. With nearly all the *paniolos* and other workers being Hawaiian or part Hawaiian, there's almost a nationalistic flavor and undercurrent on the larger spreads. Such a great number of the riders are descendants of other workers who have lived on the ranch for generations. It's been a syndrome of sons inheriting the work of their fathers, then growing into it, and in turn rearing sons to do the selfsame thing. It is little wonder, then, that the *paniolo* is so dedicated a breed of man.

Contrariwise, it has been a popular fallacy with some people that the Hawaiian is too indifferent and happy-go-lucky to be a highly satisfactory and diligent worker. In truth, the Hawaiians are not only excellent workers as cowboys, but they measure up to members of any other of the races in the Islands who turn to cattle work. The kind of Hawaiian who picked out the work of a *paniolo* for his lifetime career has proved to be a man who has a no-nonsense cut to his jib. He is not the wantless Polynesian so often, and so unfairly, depicted in literature. He is, in fact, a very special man.

In a throwback to the Old West, the vast Parker Ranch has appointed these six formen — with 289 years of ranch employment among them — to oversee the territory. True *paniolos*.

Standing, l. to r. they are Joe Pacheco, Harry Kawai, Henry Ah Sam, Willie Kaniho. Kneeling, l. to r., they are Bob Sakado and Frank Vierra.

Big Island *paniolos* on Fiesta Day being reviewed by Hawaii's State Governor, "Bill" Quinn

Ranch owner and four of his Big Island modern ranch hands. (Modern day *paniolos*)

Hands of a proud old *paniolo*

Big Island *paniolo*

Paniolo Jimmy Lindsey (82 year old cowboy of Hawaii and still active riding)

Oldtime Big Island *paniolo*, Hogan Kauwe

Charley Lindsey (Oldtime Hawaiian Cowboy)

Willie Kaniho

Modern-day *paniolo*

Paniolos on Lanai

Vaquero way of transferring cattle from shore to ship. The Mexican vaqueros taught the method to the Islanders. Today in most places the animals are run aboard from wharves.

Fighting the cattle into the sea for transport aboard the *Humuula*

Vaquero way of transferring cattle from shore to ship. They taught the method to the Islanders.

Paniolos floating cattle out to steamer at Kawaihae, Hawaii

Transporting (or swimming) cattle from shore to steamer *Humuula* at Kawaihae, Hawaii

Vaquero method of transferring cattle from shore to ship. They taught the method to the Islanders. All this predates the building of long wharves extending out to sea.

Old-time loading of Parker Ranch cattle

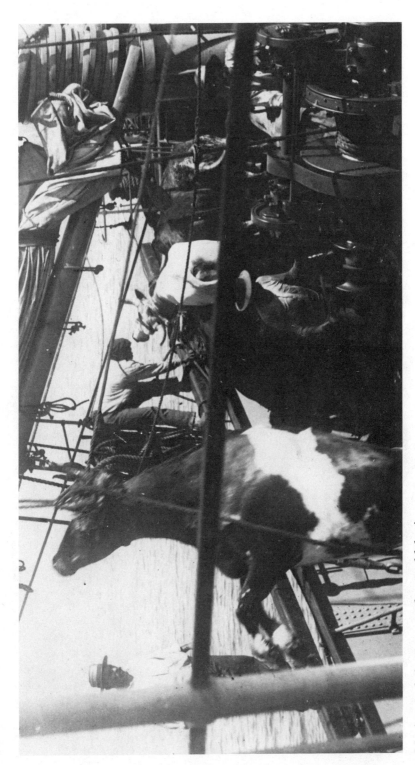

Moving cattle by sea was slow and laborious

Vaquero method of transferring cattle from shore to ship. They taught the method to the Islanders.

Vaquero way of transferring cattle from shore to ship. They taught the method to the Islanders.

Hawaiian cattle herds being transported via SS *HUMMUULA*

One of thousands of cattle roundups in Hawaii

Another one of the countless roundups on the Big Island

One of the hundreds of cattle gates on the Big Island.

One of the many Big Island calf roundups with modern *paniolos*

Modern-day *paniolo*, Alex Bell, Sr.

Branding time!

Dr. David Woo and a brace of the rare Kanaka horses which he is re-establishing in the Islands

Modern-day *paniolo*, Alex Akau

XVI

SEAGOING *PANIOLOS*

As pointed out, the Big Island was, and is, the cradle of Hawaiian cattle ranches. Island cattle raising was born there and has grown into a monumental business. And it was an incredibly crude operation in the long-ago years. One aspect of the cattle shipping there in the early days is found to be unique in that the animals had to be swum through the surf before being floated to ships for further shipment across the water to Oahu. The heavier the surf, the bigger the problem. On Hawaii, cattle were shipped in that manner from Kailua-Kona, but, for the most part, the shipments were made from Kawaihae, where the countless droves of cattle were taken aboard vessels. Back over the years on Maui, the chief ports for cattle shipping were Makena and Kahului.

The Big Island *paniolos* had a well regulated routine worked out for transferring the herds from the pastures to the port. On shipping days they would be out of their blankets shortly after midnight, have their steaming coffee, then ride to the big paddock at Puuiki where the steers had been driven the previous day. Under the cool of the night they would herd the animals to the beach, making it there by dawn, thus avoiding the heat of day where cattle lose weight under hard driving. At that point the cattle were driven into open pens where they would later be removed to the beach for transfer to the shore boats.

Kawaihae wasn't today's conception of a port at all. In fact, it didn't even have a pier or wharf until 1937. In fact, additional research comes up with a local newspaper article in 1936 which points out that a certain change would be taking place in the Big Island's hundred years of cattle shipping. It reads:

"Modernization will soon eliminate one of Hawaii's most picturesque customs — the loading of sea-going steers into boats off Kawaihae on the West Coast of Hawaii. The Territorial Board of Harbor Commissioners called for bids for construction of a concrete pier (a PWA project to cost $50,000) which will extend out over the shallow bay far enough for steamers to moor alongside and for cattle to be loaded by a telescope chute."

In truth, by the time the wharf's dedication ceremonies took place on November 17, 1937, the final cost had bloomed and blossomed to a full $80,000 — a touch of inflation, 1930's style. In any event, it brought on a new era in Island cattle shipping; an easier, faster mode.

But we're talking now about the old days when Kawaihae was wharfless and amounted to no more than a dusty, sleepy, little grouping of beach shacks, cattle pens, a limpid tidal pool, pig-styes, and a few assorted structures housing a small jail, courthouse, and post office. Along the shoreline sat canoes and fishing skiffs, while *kiawe* bushes supported drying fish nets. Offshore could be seen small craft, some under way, some anchored, and occasionally a steamer to pick up livestock and various kinds of produce.

There was little animation or sound except when the herds were being headed into the pens or being forced into the sea. At these times there would be the raucous calls and shouts of the *paniolos*, the bawling of steers, the muffled *clop* of hooves, and the upthrust of dust that gave the whole picture a beige, fuzzy cast.

From time to time the small steamers, chiefly the *Bee*, the *Hornet*, the *Hawaii*, or the *Humuula* dropped anchor offshore and waited for cattle to be brought out to them and taken aboard. Other vessels, too, were in this trade, but over the years these were the ships that were best known in the makeshift harbor. They were interisland vessels and were workhorses in their own right.

The SS *Bee*, incidentally, was wrecked on a reef off the coast of Maui in April, 1924, and had to be abandoned. She was replaced by her sister ship, the SS *Hornet*. The SS *Humuula* — a 961-tonner — went into service in 1929, and was contructed specifically for shipping cargoes of cattle. She served reliably for many years as the cattle lifeline between Kawaihae and Honolulu. However, the sting of time finally caught up with her; one day she came to a sad end when word was flashed to shore that she was in distress off the nearby island of Kahoolawe. The old, crippled workhorse was taken in tow on April 25, 1952, and hauled to Honolulu for major repairs. She never again transported a cargo of cattle. In fact, she finished her final days as sort of a broken ghost ship in waters off Florida.

The backbreaking work of herding the cattle into the water took some doing and was laced with dangers. Powerful, dependable horses of Percheron stock extraction were used in this work. Stripped saddles were cinched to them because seawater rots leather so quickly. Only the best trained horses were employed in this seagoing operation, for the cattle were always fractious and fighting-mad, or simply panicked, when being headed into the surf. So, it took a seagoing horse to push through the breakers with a man on its back and a balky steer on a rope. It meant that rider and horse had to fight each steer out to where the crew of the shore boat would grab hold of the swimming animal's horns and lash it to the side of the craft.

This was done by the seagoing *paniolo* tossing a loop over the steer's snout, then bringing the rope up the cheek and making another loop around the head. In this manner the animal's head was kept clear of the water and its breathing was not restricted.

Finally, when both rails of the shore boat were fully occupied by the lashed heads of steers — five or six to a side — the craft was propelled farther seaward to the waiting steamer. Originally this propulsion was generated simply by the crew pulling on a handline which was stretched from a deeply imbedded post on the beach to the offshore ship. In later years, of course, an outboard motor on the whaleboat supplied the propulsion.

Upon bringing the floating cattle alongside the ship, the animals, one by one, were hoisted from the water by a sling attached to the vessel's boom. It was tedious work. Shuttling back and forth with cargoes of ten to twelve steers each time took a lot of time, patience and labor. Particularly was this so when the seas were rolling and the winds were whipping.

Sharks cruised the Hawaiian waters, and there was always the danger of the big biters of the sea moving in on the cattle and horses during this crude shipping operation. Hammerheads and blue sharks abounded. Sometimes the vicious *niiuhis* (tiger sharks) come in close to shore and, particularly when they're in packs, attack about anything that moves. It had happened at Makena on Maui when the Baldwin Ranch at Haleakala was shipping cattle from that bay. It had happened at Kailua-Kona when other ranchers were shipping via water; yes, and it had also happened at Kawaihae.

So, with that hazard always facing them, the *paniolos* had to be ever alert with a watchful eye for sign of any tall, gray dorsal fins cleaving the water's surface. Cattle and horses were particularly vulnerable to the undersea marauders, because of no means of defense against the unseen attackers. But the *paniolo* took all this in stride, even though he knew too well that he himself might be the victim. One thing was certain; it wouldn't do to be accidentally unhorsed... It was just one more calculated risk in his business. As always, nature is neither kind nor cruel, just indifferent.

A reference to the old-time form of cattle shipping by sea in 1874 is found in Frank Vincent, Jr's *Through and Through the Tropics:* "After having embarked on a schooner (at Kawaihae) preparatory to returning to Honolulu, I was much amused at the manner in which some bullocks were shipped. The schooner was anchored about 200 yards from the shore. The animals were first driven into a small enclosure, situated on the beach and surrounded by a high stone wall. A rope, by means of which a large scow could be pulled to and fro, was stretched between the enclosure and the vessel. Two *kanakas* (native men) on horseback now set to work. One of them lassoed one of the cattle and twisted the lasso two or three times around the horn of his saddle while the other seized the animal's tail. Both horsemen then started for the scow and however obstreperous the bullock might have appeared at first, the logic of the lasso and the argument of the twisted tail soon proved to him the prudence of submission. It is said that the bone in a bullock's tail is easily fractured and that when thus caught, the animal will resist until fracture threatens after which he becomes quiet. Reaching the scow, the heads of cattle were secured and while two or three natives pulled the boat, the animals were compelled to swim to the schooner to which they were hoisted by means of a tackle and a broad girth around the belly. They were placed in double rows, their heads to the centre and their horns firmly lashed, and in this manner were transported to the capitol."

But now in these later years, and with a fine, serviceable pier jutting out from shore into the sea, the shipping problems are fewer. Too, today, instead of herding the animals overland from inland paddock to Kawaihae, they are transported in giant trucks, and naturally arrive there without the fatigue and loss of weight that the long drive on foot would engender. This time and labor saving method is enormously efficient and economical.

Also, today the cattle are moved across the water on large, roomy barges that are an improvement over the pitching ships that used to do the job. Neither were the cattle dehorned in those days, and the animals had to be lashed by their horns to the railings set up on deck in order to keep them from being jostled and thrown around, and consequently panicked or injured. Today the dehorned, or poled, cattle are left free, but closely packed in bulkheads that serve as small individual paddocks. Added care is furnished them by spraying them with cooling water, for exposure to sun, wind and salt spray at sea is a far yell from their normal life on the range. It doesn't take too much rough, inhumane handling to make cattle sick or put them in a wild state of franticness. That makes for poor beef.

In truth, the seagoing *paniolos* had to, and do now, know their profession equally as well as those on the "landlubber" details. Cattle at sea react somewhat differently from cattle on the range, and it's always been up to the *paniolo* to contrive and devise to meet the difference. Over the years he has measured up, and the good delivery of beef on the hoof to Honolulu now runs into the millions of tons. It is work well done by top craftsmen.

XVII

THE MODERN ASSEMBLY LINE

As the times went by, better and better imported blood stock continued to improve the Island ranch herds. Particularly were the herds on the Big Island ranches blooded up with fine Herefords. However, it appears from the records that the first importation of a Hereford bull was delivered to a Mr. Moffitt in the 1850's on Oahu. His herd at Kahuku was improved and enriched by the addition of this fine blooded animal. The Big Isle herds were improved by the legendary bulls, "Oddfellow" and "Kamiakin." Purebred Hereford bulls from Washington in 1908 contributed tremendously to the improvement of the Island's cattle strain. Some of them were the get of the famous "Beau Donald 31st." Also strong in the pedigrees of these early imports was the blood of the highly regarded sire, "Anxiety 4th." In short, bulls of this stamp were rapidly stepping up the Island herds in class.

Importations of more blooded stock was continuous, and most of the animals came from the Middle States on the mainland. The world's finest herds were to be found in that area, although some were from the Southern and Western States. Hawaii drew on the best, and before long the *paniolos* were working with well-graded, excellent mounts. Even the steers not counted as purbreds, had been up-graded into beasts with excellent conformation finish. To this day, those who are discerning in these matters and know their beef and cattle, will tell you that the Islands' herds are outstanding in quality and grade.

As indicated, much of the cattle ranching technique has altered enormously since the days when those first vaqueros came to Hawaii with their know-how. Drastic changes in ranch operations have speeded up and sharpened many facets of the business. But, basically, the work of the *paniolo* has not differed. He is working with better equipment, like linen or rayon lassos, blooded, well-trained horses, and gear that is a vast improvement over that which he had to make by hand in the old days.

A visit to the largest ranches to observe the handling of cattle is tantamount to witnessing all the precision and efficiency of an automotive assembly line in Detroit. The work is dustier, sweatier and more backbreaking, but the men attack their chores with deftness and talent that is refreshing and exciting to observe. They operate with the smoothness, fluidity and accuracy of an I.B.M. machine.

These procedures are par for the course on most of the bigger, progressive ranches: They'll run a herd of five-month-old calves in a steady stream to highly-specialized paddock crews, and at a rate of three animals a minute, four operations are performed to perfection.

First, each animal is harmlessly thrown and quickly branded with two imprints — one the initial of the ranch (or identifying brand insignia), then a numeral which indicates the year of birth.

Secondly, each animal is administered an inoculation against pinkeye and blackleg. In case any calf already shows evidence of already being infected with pinkeye, a lotion of mixed bacterin is applied directly to the eyes.

Thirdly, an identification notch is applied to each ear; one ear might be notched with a *V*-shaped indication, the other cropped with a flat edge. All this is to assist in

The Modern Assembly Line

identifying the animals when they are in deeply-brushed regions.

And lastly, the dehorning takes place and the young bulls are castrated. The horns are quickly snipped off with large, sharp clippers, and the stumps are immediately anointed with an antiseptic coagulent to ward off infection and bleeding. Too, the castrations are quickly performed and followed up with measures to antisepticize the wound and prevent bleeding.

All in all, the *paniolos* go at their work with a hurried, careful, outdoor expertise that makes one think of surgeons, nurses and orderlies doing their thing in an operating room. The herding, chuting and roping throughout the whole procedure is carried out with expert craftsmanship and efficiency. The coordination and timing between man and horse is teamwork of the highest order.

To the uninitiated, this typical cattle ranch operation might seem hectic, primitive, inhumane — and maybe even a little insane. But through the years the operation has been worked down to a fine point. It's a scene of boisterous action, with calves bawling like off-key horns and making sudden thrusts to the left or right in an effort to escape. In a fenced-in-pen, *paniolos* are wrestling with their individual chores, some on horseback, some on foot. The ones on horseback are shoving and pushing with their mounts, backing and filling to stay the ever-moving beeves.

Some of the *paniolos* are delegated to roping and tying, some to branding, some to ear-clipping, some to castrating, some to medicating. These latter are the "ground crew," and their work is as important as that of the mounted men. To an outsider, everything is hubbub and huggermugger, but the *paniolos* know what they are doing — and they're doing it right.

The precision with which the calves are lassoed and dumped quickly to the churned-up sod is made to look easy. Two irons are being operated in that one pen; one for imprinting the ranch's brand, and one to indicate the animal's age. It's rush, rush, rush — but the *luna* (foreman) knows his men are getting the most out of their every move. They are expert workmen, these *paniolos*.

XVIII

THE *PANIOLOS* PROLIFERATE

The years moved on and by the time many cattle and horses had been transferred elsewhere and were ranging the outer Islands, *paniolos* were muchly needed there, too. So, as the years loped by, increasing numbers of local cowboys worked the stock as well on Maui, Lanai, Molokai, Oahu, Kauai and Niihau. They continue to work them today.

Evidence of cattle on Maui is noted as far back as 1806 when Amasa Delano, in his *Narrative of Voyages and Travels* (Boston, 1817) told of his sailing to Lahaina. He reported:

"They had recently brought to this island, one of the bulls that Captain Vancouver landed at Owhyee (Hawaii). He had made very great destruction amongst their sugar canes and gardens, breaking into them and their cane patches and tearing them to pieces with his horns and digging them up with his feet. He would run after and frighten the natives and appeared to have disposition to do all the mischief he could, so much so that he was a pretty unwelcome guest among them.

"There was a white man at this village who told me that they had not killed any of the black cattle that Captain Vancouver brought there, and that they had multiplied very much. This agreed with what I had heard sometime previously. I understood that the bull which they had now on Mowee (Maui) was the first of the cattle that had been transported from Owhyee to any other place. I have within this year or two been told by several captains who have lately been to these islands that they have increased so much that they frequently kill them for beef."

Maui's Haleakala Ranch, the Ulupalakua Ranch, and several others have, over the years, been great spawning ground for top *paniolos* who ride their ranges the way the original ones rode theirs on the Big Isle. Like the Lindsey and Purdy men who have ridden the plains and mountains of Hawaii for well over a century, the Maui *paniolos,* too, have worked their huge acreages with the selfsame dedication, and have absorbed the Polynesian environment into their blood streams.

Even Angus MacPhee, the former champion roper of the world, managed the big cattle spread at Ulupalakua Ranch on Maui for three years. In fact, it was there where he accidentally shot off his hand, only to later turn to introducing cattle to the barren little island of Kahoolawe, about six miles offshore to the southwest. The ranching attempt failed abysmally, but the effort was heroic while it lasted.

The Maui cattlemen are strong, out-of-door men with color and talent that is not too often found in other fields of endeavor. They, like *paniolos* elsewhere, work at their daily tasks with the zeal of men stomping snakes. They, too, play hard when the holidays present themselves. On special events, they are mounted on their best horses with ringing spurs, are decked with *leis,* have garlands of flowers on their sombreros, and serenade people with lilting songs of the Islands. It's a common sight to see them on their spirited mounts and carefully guarding the guitars or ukuleles which are slung over their shoulders.

Their Maui land is very much like the Big Island where the Island cattle industry was born. Tall precipitous mountains, deep valleys and vast volcanic formations with gaping red cones make up its topography. Great segments of it are cloaked with pungent eucalyptus, Pride of India trees, camphor, breadfruit, silvery-trunked *kukui* trees, *paanini* (cactus), staghorn ferns, tall *pili* grass, and the never-ending thorny *kiawe* tangles that hide cattle so well.

Sweeping coastlines of mind-boggling beauty ring the island, and much grazeable acreage lies not far inland, so the *paniolos* are working postcard-picture country when tending their herds. There the thin, sad piping of the plover can be heard, to say nothing of the sweet, bubbling coo of doves. Seabirds like terns, gulls, frigate birds twist and turn lazily in the skies and make for long, long thoughts among the *paniolos* observing them.

And, like on some of the other Islands, the Maui cowboys occasionally have to go on wild cattle roundups into the high, thicketed regions to halt the wild bull havoc raised with the purity of registered cattle and top-grade pasturage. These sorties by the *paniolos* involve much hard riding, much fine lassoing, much adventure — and often real danger. Outlaw bulls up in the hinterlands reign like bovine emperors of all they survey and, with thrusting sharp horns, resent any intrusion into their domain. Even the indomitable Angus MacPhee, who had been for years the ace bronco-buster in Buffalo Bill's Wild West carnival, learned that roping and capturing 1500-pound wild cattle on Maui had the kiss of death in it. He saluted the Hawaiian *paniolos* who faced the challenges and made the captures.

Particularly on the high, rough slopes of Haleakala have the wild cattle sought refuge after their maraudings at the lower, lush levels. The wastelands there, with their red cones and ridges, tend to defy *paniolos* and horses. Despite the mind-challenging, incredible distances, bottomless pits and yawning volcanic crevasses, the cattlemen push through to reach the maverick animals.

Much of the topography of the region virtually defeats

passage. Hardened blobs of once-molten lava, called "volcanic bombs," litter endless acres of slopes and bottoms. The well-known and fearsome lava slag (a'a) that slices horses hooves lies on miles of flooring like so much spent shrapnel. The whole abyss of Haleakala and its outer slopes offer man and beast a hell that went wrong. Yet all of it — even its bleakness and ghostly quiet — is something to experience and ponder and accept.

Like on the Big Isle, the grade of horses bred on Maui has been incredibly high. The *paniolos'* mounts have a touch of class. Over the years the Kahului racetrack has had homebred, hometrained speedsters that have torn up the turf with amazing clockings. The same goes for the fine, rugged horses entered in the scores of Island rodeos. And the masterful homegrown *paniolos* who have mounted and mastered those bucking broncos and steers seem to be on the improve, too. As for their roping, the Maui magicians have never had to go hat in hand to cowmen elsewhere. They run second to no cowboys anywhere, anytime.

Over on the small island of Lanai, too, is a cattle operation; nothing, of course, comparable with that found on either the Big Isle or on Maui. Today Lanai is owned by the Dole Pineapple Corporation, but before the days of pineapple take-over, the whole island amounted to one single cattle ranch, operated by a lone white man and his Hawaiian cowboys.

Lanai has always been a particularly lush island, possibly due in part to the fact that for so long it has been off the regular run of Inter-Island steamers, and consequently remained much of an untrammeled spot. For years the turkeys, pheasants, deer, pigs, and sheep have had the run of the region; a paradise for man, animal and bird. The sound of centuries is in its very surf.

The cattle ranching still persists there, but, of course, on a necessarily smaller scale. This isn't to say that cattle raising is a mere sometime thing on the other islands of

Molokai, Oahu, Kauai and privately-owned little Niihau, the northernmost island of the group. But cattle raising on those islands is not to be spoken of in the same breath with talk about the Big Isle.

It is all big business today, with prosperous cattle tycoons in the driver's seat. Bankers, investors, promoters, scientists, all have a hand in it. Yet, happily in the overall picture, the *paniolos* have been, and are, the backbone of the cattle raising industry. Those men and their women come from steadfast, believing stock. They are one with nature.

Paniolos throughout all of Hawaii — particularly those of heavy Hawaiian extraction — are Polynesian enough to recognize that nature has a hard and fast regularity; a planned syndrome, if you will, with day following night, stars moving in their fixed orbits, everything ordained like the sun, rain, tides, wind, seasons, and all growing things on earth. Back down the years they explained it all with the belief that every living thing possesses a form of intelligence and operates in harmony with everything else for the benefit of the whole scheme of things.

The Hawaiians of old believed that all objects, animate and inanimate, vary somewhat from others of its kind only because it held a different type of *mana* — spirit substance. For example, they felt that the *mana* of a warrior, a *kahuna* (priest), a ruler, a fisherman, or a taro grower, was always measured by his talent or skill in his particular field.

An unlucky man or woman was supposed to have somehow lost his or her *mana*. Likewise for a defeated warrior or a very sick or dying person. They also applied this principle to inanimate things like a *pili* grass hut that burned down, or a canoe that proved to be unseaworthy and caused loss of life. They quickly disposed of these inanimate objects. They even felt that some areas or regions on land or sea had a certain *mana* which brought good fortune to those who came there. Their belief in these principles was deep and devout. So, it was common to find the old-time *paniolos* living their lives with these guide rules; yes, and the same goes for many of today's *paniolos* as well. They can sometimes be heard to say that their horse or lariat either has or has not its *mana*.

XIX

ROBERT G. "BOY" VON TEMPSKY

But getting back to the Maui cattle ranches per se, they flourish on that island to this day, and their futures shine with the promise of even more golden tomorrows. The sad, sweet whistle of the plover is still heard there on the ranges, the *paniolos* still ride their mounts through tall grass, and great herds of red cattle continue to low and snuff and blow as they lumber along under the long white hackles of clouds coming in on the tradewinds. Ranch personnel alters, but the ranches themselves are as changeless as a monastery.

The Maui land is still a poem — as it always was. *Paniolos* continue to ride and listen to the song of life there. The flight of wild birds in stainless skies is still to be seen, and valley after valley is as green as springtime clover. And for contrast, in other areas beyond the rim of Mount Haleakala, are great romantic cone-tipped mountains diminishing into the distance.

Others now take the place of the legendary Louis von Tempsky who, for so long, worked and managed the vast Haleakala (House of the Sun) Ranch with its colorful *paniolos*. His ghost still rides those ranges. He left a mark for other cattlemen to shoot at, and it takes some doing.

Large and small, the cattle spreads continue to operate there on the Valley Isle. There are no larger nor more historical ranches on Maui than the Haleakala Ranch or the Ulupalakua (Ripe Breadfruit of the Gods) Ranch — the latter a particularly prosperous one now under the ownership of Pardee Erdman who has controlled the fine lands of Kipahulu and the State lands of Kahikinui.

Instead of being herded to the port in the old way, the cattle are now being trucked direct from the ranges to Kahului and barged from there to the Hawaii Meat Company's slaughterhouse on Oahu, or to their yards at Ewa Beach for finishing. In fact, no cattle have been shipped from Makena for many years. As for locally consumed beef, the slaughtering is done at Maui's DeCoite's slaughtershouse. And again, all of the operations still depend basically upon the very spine of the industry — the lone, mounted *paniolos* with their hard riding, herding cutting, branding, castrating, and all the other sweating-out chores that make up their life.

More than a mere touch of the old, old *paniolo* days can be found in having a warm, long talk with one of the last to remember what they were like. He is Robert G. "Boy" von Tempsky, now in his late seventies, although you would certainly not think of him as being on the periphery of life. He seems to have found how to keep time at bay. He has great charismatic powers. He's tall as a first-baseman, heavily-chested, silver-haired, and as leathery as an out-of-doors man can be. He has the face and build of a well-trained heavyweight — and the soul of a poet. For that matter, he loves to recite poetry, and clamped this writer's interest with many a stirring ode. And, best of all, he is loaded with a merry, laughing way that can charm the birds off the trees. He is a gentleman when the breed is going out of style. The late fabulous Louis von Tempsky was his uncle.

Lovely trees grace much of the area around his ranch

dwelling; tall, pungent eucalyptus trees, Italian and Monterey cypresses, dark green ironwoods that look like old *kahunas* with shaggy robes. And beyond them lie the high, rolling mountains that seem to be ever on the march and taking their age-old secrets with them. And, with the mountains forcing the warmer winds to blow high into the colder air, condensation makes for billowing clouds that you wouldn't believe. Great powdery formations swell and move in the upper regions like giant fluffs of cotton.

Maui cattle raising sounds a lot like that of Big Island ranching when one talks to pioneer "Boy" von Tempsky. In his interview with Author Ed Sheehan for the *Honolulu Magazine*, he told of some of the trials and tribulations of his times. He was quoted as saying: "When I was grown, in the 20's and early 30's — long before the road was built — we used to drive the cattle from the Hana and Kaupo areas all the way to Ulupalakua. We had to time the drives for the full of the moon. We couldn't move during the day. It was hot along that coast and the sun beat down fiercely... We'd start at night and pray for clear moonlit nights, hoping we'd have them all along the way. We'd drive about 300 head at a time and needed about 20 *paniolos* to keep the cattle single file along the narrow trail... The old cook and his mules and wagon would move ahead of us to where he'd be ready for chow in the morning... We had to pack in everything we needed — water, everything.

"It was a three-day trip, at least, and hard work. Many times I made that ride — many, many times... Ed Baldwin, too. Ed never lacked to be with his men. He never asked them to do something he wouldn't do... It was mightily rough on the cattle, too. They'd been raised in nice soft, spongy pastures and their hooves got soft. Then we had to walk 'em over a jagged rock trail for three days, and at the end of it their hooves were all worn down, many of 'em bleeding... It sure was a good feeling all around when those drives were over.

"Now there's talk of a fancy new highway to change things even more... And Ed, Chu, Asa — all those boys are gone... So many others, too....

"When I was a youngster, the food was very simple. Meat, corn and potatoes, mostly, and a little cabbage. We had none of those fancy vegetables like broccoli and that stuff. Ate the old horse corn when it was young and tender... Had a lot of milk and butter, too, but never much fish or *poi* this high up... No refrigeration, of course... Our beef was killed by old Apo, the butcher in Makawao. He'd slaughter in the afternoon, cut it up at night, then the next day load it in saddlebags on mules and deliver it all, dripping through the Kula district... There was none of that filet mignon stuff, just plain red meat in chunks.

"When it got a little sour, we'd hang it up in the shade of a tree... When the flies started buzzin' 'round, we figured they were smarter than we were, so we'd take it in and sear it a little. It was good, tasty meat, but nothing fancy. We never heard of inspections... just ate it and no one ever died from it.

"There was plenty of wild pig, specially in the cactus areas. They like the cactus fruit and *kiawe* beans — and taro and guava, too. If they forage on that stuff they taste good — nice, sweet lean meat... But if you get a pig that's been eating the young *hapuu* fern shoots, it tastes strong, not good. You can smell that *hapuu* when you cook them... can't get rid of it in a pig for six months..."

The gentleman *paniolo* of yesteryear was going a long way back, enjoying his sentimental binge and speaking low with a nostalgic note in his voice. Sheehan was revelling in the man's harking back. The words tumbled on.

"The first cowboys — the *paniolos* — were brought from Mexico by the Hawaiian government to teach the local men how to rope and tie and the rest. They sent two here to Maui, four to the Big Island and two to both Kauai and Oahu... Of course, that was a long time ago... But the Hawaiians became great horsemen. Eben Low was the one who got some of the better men to go to the Wyoming Frontier Day rodeo in 1908. There was Ikua Purdy — he was Hawaiian-Irish — and Archie Kaua and Sam Spencer. They took three first prizes from all those mainland cowboys — and this was in days when they worked big cattle, hog-tying grown steers, throwing and busting 'em. Ikua took first, San second and Archie third. Ikua held the record for two to three years until Angus broke it. Eben Low brought Angus back to the Islands with him.

"Cattle here range up to around 7,000 feet. Above 8,000 it gets pretty wild. No pigs up there, but plenty of goats. Used to be any number of wild turkeys in the old days but the roads going in drove 'em out... still a lot of chukkar in the uplands; California quail and ringneck pheasant... I see the Hawaiian owl now and then, but we have no hawks here. Sometime back they released a Texas turkey that's doing well on the lower lands... The wife and I saw a flock of about 40 not long ago. The farmers sure don't like 'em... People rent the farmlands from us and there's some 400 acres in tomatoes and potatoes now... divided up in tracts from five acres to 40, 50 or 60... We get good tomatoes if the winter months are mild.

"Have I seen that wild country 'way up? Only from an airplane a few times and from right here. There are places in these mountains where no man has ever gone..."

The fine old gentleman talks like that, then suddenly goes quiet as though he wants to hug his memories to himself. All you know is that he comes across as a big, warm man whom you want to believe. You'd feel somewhat shabby if you didn't. He's typical of the early *haole* pioneers to these Islands.

XX

"DUKE" — THE WILD BLACK BULL

The "Valley Isle" of Maui has more than its share of *paniolo* anecdotes to give to the world. And one of the richer storehouses of these bits of cowboy history is none other than this same "Boy" von Tempsky, one of the scions of the legendary von Tempsky clan. He tells the stories with color. One of his favorite anecdotes has to do with Ikua Purdy, who finally moved from the Big Isle to Maui. His story is this:

Ikua's reputation for fearlessness and talent in bringing wild cattle down from the lava-strewn highlands was known far and wide among cattlemen. For the most part, the wild cattle were steers that had once been branded and even ear-marked long before they deserted the ranch herds and took off for the high, most inaccessible mountain regions.

Word was around that a choice broken herd of wild cattle was ranging high above the Ulupalakua Ranch. It was a never-never area clear up on the rocky slopes of Haleakala Crater — high, cold lands where the wind drives the mist like fleeing ghosts; an unlikely region for cattle. It was reported that one outsized black bull reigned supreme over that group of stragglers. Several sightings had been made by lone range riders, and they had even identified the old branding. Rumors kept drifting in that this lead bull was an incredibly outstanding beast; a deep-chested, heavily-boned brute, black as the king of clubs, and majestic in his walk.

The rumors persisted for long and long, and whether or not Ikua Purdy thought of the animal as having a great potential for stud use, it is not known. Possibly it might even have been mere curiosity on Ikua's part that finally prompted him to form a roundup gang to comb the upper heights and bring down the black bull — if they could locate it . . .

It took the band of *paniolos* under Ikua's guidance a good deal of crunching, grunting riding up through *kiawe* and cactus-tangled regions to even get into the jagged *a'a* areas that were so murderous to the horses' hooves. The riders' mounts were pretty well blown from the unending climb, and the rarefied air at that above-7000-foot-level only compounded the breathing problem. Other wild cattle had been sighted, but not the wanted bull. Purdy's plan was to locate the black bull with its own herd, and then to work the whole group back down to the lowlands where the corrals were located. He and his cowhands had done this in the past with wild cattle, so everything appeared feasible and right. A calculated risk, to be sure, but probably the correct move.

However, Ikua Purdy had not reckoned with this kind of a wild bull, nor with the kind of wild cows which the king of the rangelands had selected for his harem. When the *paniolos* finally located the big black bull with his private herd high and deep among silvery-trunked *kukui* trees, they were soon to learn that this huge bull had gathered only outstanding cows around him; it was as though this bovine monarch would have nothing to do with the run-of-the-mill cows.

As the *paniolos* moved in for the roundup, those closest could see the giant bull paw at the ground, snort and turn

"Duke" – The Wild Black Bull

in all directions as though dementedly anxious to take on any man or horse that would dare invade his domain. The wide horns raked the air in warning and contempt. Even Ikua, with all his disdain for danger, could feel the air charged with the electric fierceness of this very special wild bull.

With the yelling and shouting of the *paniolos*, the cows suddenly broke enmasse from the thickets and headed for lower land as though taking the line of least resistance. At least it was downhill and the steep slopes offered faster escape. The enormous bull raised his head high, snorted in either disgust or surprise, for it was obvious that his harem was being driven from him. It was even to Ikua's astonishment that the bull took off to rejoin the herd. Defending and holding what belonged to the bull was apparently the animal's natural instinct. The herd leader thundered in the directon of the cows as though to halt their flight or to lead them to safety. There was no way for man to know a bovine's thinking in a tight case like this.

But it has always been said that Ikua Purdy "thought like a cow." Many had said with admiration, "Ikua has a smart cow's head." And the speaker was prideful that he could count Ikua Purdy as a personal friend. Ikua was a man for all seasons. Those riding with him now on this wild cattle sortie, too, were in league with him in his every move, his every plan.

So it was that the roundup started working at a gallop downhill toward the distant corrals. There was the usual plunging and lunging that accompanies steep, downhill riding. The intermittent yells of the *paniolos* rent the air. The black bull was now at the head of the herd, running hard and apparently trying to show the way to escape. The route continued with pounding hooves, hoarse cries from the riders, and the heavy panting of the animals. The rustle and swish of staghorn ferns, plus the crackle of camphor, *kiawe* and lantana tangles all added to the pandemonium.

Ikua Purdy counted this as the most nimble and fleet

wild herd he had ever encountered. The black bull had selected his herd of cows with an instinct that was uncanny. Sturdy, fleet-footed animals, every one. They rumbled downward enmasse like a solid, spiked machine that would be unstoppable.

For miles this mad roundup continued avalanching down through terrain that was even beginning to slow down and fag the stout range horses. Some of the unbelievable thickets that the herd plunged through had the riders fighting hard to hold to their saddles. Thorned branches clawed at them, trying to dismount them or shred their clothing. This was surely a wild herd that could escape back to the highlands if it could once get far enough ahead to turn back and disappear into the tangles that laced so many of the nearby canyons. So, the headlong downhill riot of man and beast thundered to and past the holding pen that had been erected to keep the wild cattle captive until domesticated by the presence of tame cattle.

The whole herd ended up on the beach west of Nuu, and for the cattle it was the world's end with no place further to go. A big surf pounded there and the panicked herd turned with lowered horns and heaving sides. Then, confused and in disarray, they started milling, searching for a way out. All milled — all but one; and that one was the big black bull leader. He headed through the surf, apparently trying to lead the rest to freedom somewhere out on that blue horizon. None of the cows stampeded out with him.

Even Ikua Purdy sat his horse there, transfixed and staring. This scene made no sense at all to him; a new experience despite all his years of wrangling on the range. His old eyes had seen many strange things in cattle behaviour, but never anything to match this.

It somehow seemed unlikely that the bull would be able to thrust through the walls of white combers that roared in to explode on shore. The other *paniolos*, too, sat with mouths ajar and watching the progress of the wild bull. The animal did breast the first tumbling comber, wavered and rocked from the impact for an instant as the weight of the water jammed it back. But only for an instant; the bull was then again churning with all four legs and shoving through. Now it was beyond its depth and thrashing hard. It rammed its head and horns into additional oncoming breakers, kept swimming and churning with incredible power. Farther seaward the swells were racing shoreward like long blue garden walls. They towered and glistened in the sun, and it appeared that no living thing could possibly negotiate them. Their raw strength would not be denied.

For a few long minutes — unbelieving minutes — the *paniolos* were oblivious to the rest of the herd onshore. Their undivided attention was riveted on the fast-disappearing black form that was on a seaward route with apparently no intention of turning back. Excitedly they watched, in awe they pointed, and noisily they gibbered to each other about this strange phenomenon that was taking place before their very eyes.

When the distant animal finally made a slight slant toward the east, Ikua simply said in wonderment, "Swims like Duke Kahanamoku!" It was the finest verbal orchid he could have paid the seaborne beast, for at that time the great Hawaiian swimming champion was at the zenith of his aquatic career. Also, the great Olympic champion, Kahanamoku, was a man who left fragments of himself with everyone who knew him.

"Y-yeah," panted one of the riders in admiration. "But this Duke'll be shark bait before long..."

"Right," agreed another cowman. "There's no other beach for a helluva bunch of miles."

And it was true; so, the consensus was that they had seen the last of that outstanding, fearless animal.

Regretfully and a little sadly, Ikua and his men turned their attention back to salvaging the remainder of the wild herd stranded on the beach. But none of them could quite get the lost bull out of their minds. Their sporadic gazes seaward told of their surprise and awe. And respect. The disappearing bull had left a vacuum for all of them.

Long after the balance of the wild herd had been corraled, branded and ear-clipped, the *paniolos* continued to shake their heads in astonishment and comment on how the big "Duke" had escaped them and swum to sea.

Several days elapsed with the *paniolos* still occasionally mentioning their experience of having seen the black bull swimming strongly out of sight to sea. All of them spoke of the animal as "Duke" — and with something of reverence. There was much guessing and hazarding as to how the beast finally came to its end at sea. Had it been shredded and devoured by the ever-ravenous sharks? Had sheer exhaustion halted the animal's struggling and reduced it to a drowned cadaver? Or had it tried to swim in somewhere at the base of the sheer cliffs and been battered to death by the mountainous breakers that blasted there with thunderous impact?

So, it was a thing of incredibility one day when one of the ranch's water trough checkers came riding up to headquarters to get a rifle for a special shooting purpose. He had discovered the black bull hopelessly trapped in a small boxlike cove a few miles from where it had originally entered the sea. With the precipitous cliffs making all escape impossible for the bull, the humane thing seemed to be to give the animal a quick merciful death in lieu of allowing it to die slowly from thirst and starvation. The *paniolos* at headquarters could hardly grasp the fact that the bull was still alive.

The rider took his leisure in getting the rifle, then rode away slowly, for time was surely not of the essence in this instance. He obviously was trying to think of some other alternative, and even muttered that the poor beast deserved a better fate. For that matter, he stopped several times en route to tell the story of his strange finding to incoming fence riders. Their faces clouded at hearing the news.

Then to compound the mystery, the water trough checker finally made it back to the top of the perpendicular cliffs — and you guessed it... The bull was nowhere in sight in that deep, barren hollow. Looking hurriedly and perplexedly into the sun, and scanning the offshore waters, the rider could see no evidence of a swimming steer. The week-long storm was still hanging on; in fact, judging by the exploding surf against the cliffs, it had increased. Finally, he spotted the glint of sun on horns far seaward and way to his left. He crossed himself in respect for the game animal that was, for the second time, breasting heavy seas with scant chance of survival.

Upon the *paniolo's* return to headquarters, his story about the bull's second foray into the storm-torn waters only added to everyone's reverence for the beast. There was much talk about animal courage, animal quirks, and animal prowess. But, most of all, there was palaver about what a great stud bull this one could be were it to be domesticated and added to their ranch herd.

From time to time, all this made for lively conversation among the cowhands for months on end. The legend of the animal was in the making. It continued throughout the year, growing and expanding like all legends are wont to do. Black "Duke" became something of a byword among the *paniolos*.

Then to add luster to the memory of that fabulous black bull, Ikua Purdy and some of the *paniolos* actually spotted the selfsame animal high in the mountains several months later. It was unmistakably the identical stray, for they again identified it by the mark of the branding iron, plus the same ear-clipping. Most of them were for again trying to capture the mammoth beast, but Ikua called them off with, "Hell, leave 'im be. That's big 'Duke.' A bull *that* brave deserves his freedom!"

Ikua's word sufficed. No one made a move to ride herd on the amazing animal. They nodded in agreement to the decision. To the man, they wondered and discussed just where and how the wild bull had ever managed to scramble up onto some beach that long-ago day. They were still remembering the sea's towering breakers, the rocks, the shark water, the everything that could reduce a live bull to so much inert flesh, bone and hide. Duke's escape had been a modern miracle, and they spent much time hazarding guesses. Too, the bull was seen several more times after that, high in the remote spots; but his freedom had been made inviolate by old Ikua Purdy's declaration to let the beautiful brute roam at random for the rest of his natural life. The *paniolos* find it good to know that old "Duke" lived his final days high in the mountains where the winds are tinged with eternity and where the purple haze hangs over the hills.

"Duke," he is still called in memory — even though he has long since passed off this sphere, just as has his famous namesake who also left us back in 1968.

XXI

CATTLELAND COURAGE

Wiry, courageous and capable, he was one Oriental who knew his cattle the way you know the back of your hand. Ah Sing worked the boats and barges that hauled the animals from the outer Islands to Honolulu. He knew his way around in cattle shipping by sea. And when he wasn't working with seaborne cattle, he was herding them, roping them, branding them, et cetera on the Island ranches. He was no Lindsey, and certainly no Purdy, yet a good, responsible man. A "roustabout" *paniolo*, some called him — but a monumentally reliable one.

Rancher Robert "Boy" von Tempsky of Maui still talks about the time Ah Sing was on the cowboy gang at the big Ulupalakua Ranch of Maui, and the hard-riding Oriental showed a grittiness for carrying through that is seldom seen these days in most fields. In fact, the incident even prompted the legendary Ikua Purdy to remark in awe, "That goddam *pupule pake* (crazy Chinese) he got no *makau* (fear) in his body!"

To quote von Tempsky: "We were driving steers for slaughter in a very heavily-covered *ekoa* and Guinea grass area in the Kahaahena section. Some of the older *ekoa* had grown nearly to a small-tree size about six inches in diameter. In this pasture a wild bull — a really huge animal — would drive along with the rest of the animals until they came to a certain spot, then he would break away and take a good number of steers with him. The bull had done this several times in the past, and it was uncanny how the animal accomplished it. Several chases by hard-riding *paniolos* had netted them zilch. It was that kind of country — terrain that you wouldn't believe.

"The order before the drive was to try to rope him if he broke. A pin bullock (a tamed bullock used for calming a wild steer) was taken along for the purpose of tying it to the stag in case they were able to capture the latter. There appeared to be two chances of such capture — slim and none. Things seemed to be working out as the riders went on shoving this herd of cattle down a long slope toward a distant corral. Hopes were Himalayan high.

"Sure enough, the huge stag was in among the herd, looking jumbo-big and raw with power. As it crunched along it appeared to be working toward the fringe of the herd in order to break away and lead some steers into heavy thickets to the right. In its progress, the stag was cutting a group from the herd about the way a trained cutting horse would do it. It was something to see, and it made the *paniolos* blink.

"Ah Sing, closest to the stag, quickly swerved his horse alongside the escaping steers, reined in hard and cut them back into the original herd formation. But the stag was free of the herd now and racing for the *ekoa* stands. The balance of the riders continued on with the galloping herd, but all of them had had one last look at Ah Sing lifting his rope from the pommel and bent upon trying to get a loop over that bounding wild stag. The chances looked slight, for the animal was a blur of speed and the thickets close at hand. Stag and horseman disappeared into the tangles the way a runaway freight train slams into a tunnel and vanishes.

"Approximately an hour later when the cowmen had

all the steers properly corraled and quieted down, they looked up to see Ah Sing's sorrel horse come trotting in riderless and blown. In addition, it was bleeding from cuts on its flanks and chest. Even the saddle was gone. The *paniolos* gaped in wonder at the specter of it. Quickly, two of the men hurried to the injured horse to take it in hand and tend to its wounds. Without a word, the balance of them mounted their horses and spurred them in the direction of where they had last seen the disappearing Ah Sing.

"After much combing and ransacking of heavily-brushed acreage — areas tangled with thorny *kiawe, ekoa* and cactus plants — the riders came upon a scene that left them boggle-eyed. There was Ah Sing, his face and hands bleeding from cuts, but still sitting in his saddle with the fork jammed up against a stout six-inch *ekoa* bush. His rope was still cinched to the pommel, and at the other end of the line was the giant stag lassoed and captive. To add to their amazement, the Oriental was nonchalantly rolling a Bull Durham cigarette. Their silence told of their awe, and the stag's thrashing told of how well the noose had been thrown and secured.

"Not until Ah Sing mustered a smug smile, lighted his cigarette and blew a cloud of smoke, did the *paniolos* start their excited questioning. To them the whole situation presented a mind-bending dilemma. They were all talking at one time.

"Ah Sing broke in upon them with, 'I no let him go this time!'

When the Oriental was found not to be seriously hurt, the back-slapping and compliments began. It was his day. After the stag was slaughtered, it dressed out over 950 pounds, and Ah Sing had the prestige and joy of being the donor of much top grade beef to his pals. Unfortunately, he was a vegetarian himself...

This Ah Sing *paniolo* never reached the heights of stardom that some of his cohorts did in the way of cowpoking, but his life on the Island ranges often departed from the humdrum or the ordinary. Robert "Boy" von Tempsky likes to tell of the time when the Oriental cowhand did a downhill sprint that probably fractured all existing 100-yard dash records of the past century. Unfortunately, there was no clocker or measured distance, but von Tempsky is positive that a record was established — even if unofficial. In his own words, he tells us:

"The upper reaches of the Ulupalakua Ranch were at one time covered with a lush plant called *paa makani*. This herbage grew up to six feet in height in rich pockets of land there and was the favorite grazing for semi-wild and wild cattle. At the time I have in mind, there must have been more than a thousand wild and semi-wild cattle scattered in the region — a rock-strewn country apparently spewed up in the last days of fiery creation. We seldom worked that area because of the almost impossible volcanic terrain.

"We wanted the cattle that roamed that cold, windy region, but trying to round them up and herd them offered so many obstacles. The air at that elevation seemed to brittle one's bones. In time, it developed that the most practical way to capture the cattle was through the use of the few water holes that lay in an isolated section. First, we fenced in the areas where their main watering holes were located. Then we erected gates on the main well-traveled cattle trails that led into them. For many weeks nothing was done to try to trap the animals, and in time the open gates on the main trails no longer spooked them. We would wait for the dry season to hit when there would be no dew on the grass, and water remained only in the main water holes. Meantime, this big trap was connected up to a long, wide wire chute so we could funnel the beasts to a corral where we could handle them.

"Finally, when the blazing heat set in and the wild cattle had become adjusted to passing through the open gates, we were all set for our next move. We waited until the parched cattle had gone through the open gates to reach their few remaining water holes. Everyone had had his assignment to close the gates at midnight, so everything shaped up as foolproof. We had all the ranch hands, including the cook, bedded down at our 6000-foot elevation *puuolai* (volcanic cone) mountain shack. The place was tomb-silent and the air was still heavy with the summer's drought. But we were content with the thought that the cattle were gathered around the shrunken water holes like fruit flies.

"At dawn we were in our saddles and moving toward the fenced-in area where better than a hundred wild steers were milling around frantically. They were trying to find an opening, and the thunder of their hooves told us that they weren't going to be easily stopped.

"We had spread out when Ah Sing suddenly saw the whole herd pouring toward the one gate that had accidentally been left open. Several panicked steers had already found their freedom there. Ah Sing spurred his horse in the direction of the open gate to try and close it in time. The grade was strictly downhill and steep. The rest of us were too far back to reach the gate before all the cattle would rumble through it, so we were only witnesses to what happened.

"Ah Sing was within fifty yards of the open gate when his plunging horse stumbled on the steel-hard lava, somersaulted and broke its neck. Spectacularly, the spill hurled Ah Sing clear, and he landed on his feet on a dead run. He was obviously terribly shaken, for he looked like a drunk as his legs churned and his boots thumped on the lava surface. Without a break in his wobbly stride, he staggered on to the gate and swung it closed just an instant before the main herd of wild cattle pounded up to it.

"The rest of us got to him and we wanted to stand up and cheer. We babbled inanely about how it had all come about, and no questions were asked as to who to blame for the gate having been left open. Ah Sing was a little in awe of his narrow escape. He could hardly believe that he had miraculously turned back the stampeding wild cattle at the last splinter of a second. The loss of his horse was a real trauma for him, but he realized that he was lucky to have survived the tragedy of his mount's fall.

"One and all, the *paniolos* again saluted him for his outstanding feat as a cowpuncher. If they had their way, he would have been awarded the Silver Star for gallantry in action. At this rate it looked as though he might yet become a legend in his own time. In their minds it couldn't happen to a braver guy.

Maui had its courageous girl characters, too, who stood up to trying circumstances just as well — and maybe even better than many of the male riders. Cattleman von Tempsky likes to tell of the time when a wild-horse-roping party went sour on the slopes of Haleakala. Sour? In truth, it almost went tragic!

He tells us: "A band of wild horses used to roam in the *mauka* (mountain) land of Erehwon, and every summer all the kids returning from school on the mainland would make up roping parties and try to lasso one of those rangy speedsters. The wild horses wandered the 8000-foot elevation where it was cold and the terrain was unbelievably rugged. Those tough mustangs had lung power that was incredible, and could outdo our best horseflesh.

"The party on this occasion consisted of Mr. Frank Baldwin, cattleman, and his three sons, Edward, Asa and Chu, plus my cousins, Lorna and Erol von Tempsky.

"No wild horses were located for many hours, but the group was determined not to be wholly outdone in the lassoing of something. They finally spotted this six-year-old wild bull that had never felt a noose around its neck, and they thought they would try their luck. With the group working excitedly from all sides, Chu was the first to get a line on the big animal. Unfortunately, he caught only one of the bull's front quarters in the loop. Now you've got to understand that all these kids had often been warned to never get below a bull, for hell would likely start hitting them. But excitement being what it is, kids are apt to forget what they've been told.

"The bull began going berserk, and regrettably in the mad scramble, Lorna got below the enraged animal and right in line with its downward plunge. Chu fought hard to stay the bull's charge and tried to keep his lariat taut. But it was no go! The security of Lorna and her horse was not worth a whistle in a hurricane. The bull was section boss this day and roared down the slope to catch her horse before it could escape the terrible impact of the lifting horns. Lorna's mount was ripped open from forequarter to hindquarter. Entrails bulged from the horse's belly as it reeled and staggered a full hundred feet and collapsed like a house of cards. The bull rumbled on into the underbrush, dragging a short length of the broken lasso. Lorna had been thrown clear and wound up with painful scrapes and bruises. The kids were more than a little scared.

"But Lorna wasn't thinking of herself. All she was conscious of was the pitiable state of her horse. No one in the group carried either a gun or a hunting knife. Lorna winced at the sight of the suffering animal as it lay prone and quivering. Anguished, she knew she had to do something and do it fast. She took her courage in her hands and, with only the aid of a pen-knife, she cut through the tough hide of her fatally wounded pet mount and severed its spinal cord, allowing it to die instantly.

"The fact that the rampant bull was tied down and killed the following day by the *paniolos* didn't impress the kids nearly as much as did the memory of young Lorna bravely dispatching her favorite horse in the most humanely way possible. Even the *paniolos*, when they got word of the incident, saluted her for her fast thinking and uncommon courage. Imagine — a young girl like that! Well, courage is where you find it..."

XXII

PANIOLO'S COWTOWN

The *paniolos* on Hawaii have their chief meeting place, and it's the unique town of Waimea — or Kamuela, as it is also known. Kamuela means "Samuel" in Hawaiian, and it's the official post office designation for the place — and the names are interchangeable. People who know a little about the history of the region ask if the town was really named after Samuel Parker, one-time owner of the immense Parker Ranch. Others, with part-knowledge, inquire if it's true that it was named after the former postmaster who answered to the name of Samuel. You'll get many arguments, pro and con on this.

But that's neither here nor there; the point is that Waimea/Kamuela was originally a village modeled after a *paniolo's* needs. The little town sits at the base of 13,796-foot Mauna Kea. In winter the high slopes of the mountain are blue-deathly cold in shadows, and even in summer and spring where the sun illuminates them they gleam coldly. When one is up there in the heights, it's a stillness place, with only the distant sea moving and sparkling.

The *paniolos* had need for a general store where they and their families could purchase supplies and merchandise. The large ranch restaurant was important to them;

and so was the general store with its wide variety of goods. These were the important facets of the town for them when it was no more than a tiny village, empty and dozing in the sun.

Now that it has escalated into a town of some prideful size — what with the advent of the automobile and truck — there have been the garages, the gas stations, and all the businesses that service the car owner. It has its own tidy landing field where jet planes shuttle on and off like incoming and outgoing starlings. Too, places of worship have sprung up — Christian churches, synagogues, Buddhist temples, Catholic, et al. Many other transformations have taken place and the town's image has been completely turned round. Banks and markets now shine there with burnished fronts and sparkling interiors. When it was just a dusty cowtown with a lazy, but happy, cloak of *manana* enveloping it, it was simply spoken of as picturesque. Today it is picturesque in a new and distinguished way. It has a cathedral serenity, and many people go there just for a slice of life. The burgeoning tourist trade is beginning to find it. On a clear day there you can see forever — and most days are utterly clear.

The little cowtown is now wearing clothes of a gay and sprightly color. It has a strong leaning toward architecture of the Victorian Period; it is also the Monarchy Period which dates back to the King Kalakaua days. The decor of many of the buildings reflects that era in both construction and color, and it adds a touch of champagne to the air. With this tasteful royalty motif predominating at many sites, it gives rise to a nostalgia in connection with those long-ago days when the *paniolos* were first coming into their own. The late 1830's, '40's, '50's and on, were Monarchy days; clear up until 1893 it was a Monarchy, and when the reign was abolished something wholly Hawaiian went away with it.

But the *paniolos* remained, doing their jobs — even bigger jobs — for the beef markets, the cattle herds, and ranch acreages, were all escalating. Everything was on the increase; expansion everywhere. The Hawaiian cowboys still needed their cowtown, so it grew along with their trade. It is there today, lusty, colorful and a haven for *paniolos* and their families in from the various ranches.

Mister *Paniolo* — and he certainly deserves that "Mister" appellation because of the outstanding job he has done over the years — enjoys his siesta in his cowtown just as his forefathers did in bygone years. He is seen there often, a windburned, rugged individual who walks with a swinging gait and slightly bowed legs. He walks like a man who knows where he's going; and he *does* know, for his record back of him is good and sound. He's prideful about it in a quiet, modest way.

But he's best and he's happiest when out on his wide range, tending the herds under skies where clouds scud overhead like great white sailing ships. He likes his paddocks, the snort of his fine-blooded horse beneath him, the creak and smell of sweat-laden leather, the sodden *clop-clop* of steers' hooves, and the whisper of his rope as it snakes out true to the target. He prefers all these *paniolo* things — and he finds the cowtown really is a great place to visit and talk *paniolo* talk with others of his kind. The *paniolo* is still around, and he will continue to be around. You should meet him; you'd be the richer for it.

XXIII

FRANK BERNARD VIERRA

A solid book could be filled with individual profiles and tales of the old-time *paniolos* who have given so much to Hawaii's cattle empire. As pointed out, the *paniolos* are scattered now over all of the Islands, for cattle ranching prevails to a greater or lesser degree on all of them. But to portray the typical Hawaiian cowboy, here are the profiles of several who are products of the Big Isle, and they depict the manner of men that have made Hawaiian ranching what it is in all Hawaii.

Some were related directly to those first vaqueros who came over from California circa 1830. Most were not, but either they or their forefathers, through exposure to the vaqueros, soaked up the latters' technical skills and general know-how in handling cattle and horses.

Big leathery Frank Bernard Vierra rounded out a full fifty years in June of 1970 on the Big Island. Still in good health and as rawboned as a brewery horse, he grinned at finally turning himself out to pasture. Despite the grin, upon his face and body was the undeniable patina of a lifetime of toil.

He was born in Waimea, and at fifteen was a full-time worker under his father on the ranch. Like so many of the other old-time *paniolos*, he worked at many chores before he became a full-fledged cowboy. This involved planting robuster trees for windbreaks on the ranch, maintaining pipe lines on the range paddocks, dam building, breaking horses in the breaking pen, training mules, etc. By 1932 he was qualified to join the cowboy gang, and he started under the top teaching of Willie Kaniho, Sr. For thirty-one years he worked as a *paniolo* — and a top one he was.

Although of Portuguese extraction, he spoke no word of the language. In fact, living and working with Hawaiians for the most part, he was fluent in the native language and spoke it most of his waking hours. He sang the old Hawaiian cowboy ballads with the finest of Hawaiian interpretation. Right up until his retirement in the summer of 1970, he was still in the midst of the toughest of *paniolo* work like herding and branding, and his roping was still first-cabin stuff. And his love for cowpoking was still Tibetan high.

He was asked what his fondest recollections were of all his years as a *paniolo*, and he answered: "Shipping cattle at Kawaihae. We used to *hoau* (swim) the cattle out to the boat. I remember New Year's Day many years ago when we had horse races and the big *luau* (feast) that followed it. The good Hawaiian spirit was very much there and the next day we packed our blankets and left for Keanakolu high on Mauna Kea to separate cattle. For our food we packed salt meat and *ai paa* (hard poi). That's the best — can't beat it. Our pay was small and we worked hard, but our benefits were good and everybody was happy."

And that was Frank Vierra all over; he gave so much and asked for so little. He was in the Indian summer of his life, but he still looked tough enough to scratch matches on. He left the ranch with his heart full of aloha and fond memories of a lifetime as a *paniolo* there. Happy, yes — although he knew he'd often be lonely for the days he had known...

XXIV

JOE PACHECO

Another legendary *paniolo* who served out a half century (forty-nine years to be exact) of dedicated ranch work is Joe Pacheco, who retired on the Big Island on July 4th, 1968. Big, swarthy and capable, he could still rope with the best of them. Those were unbroken years he served on Hawaii's vastest ranch lands, and he, too, was a *paniolo* who worked at a multitude of jobs before developing into the top-flight *paniolo* which he became.

He was a foreman on the day of retirement, and chortled in retrospect when relating that his first assignment was to help fight the famous Hanaipoe fire which burned and smouldered for six months in 1918. The blaze was started in Umikoa and spread all the way to Hanaipoe. He laughed in telling that he was the only "kid" among the men who were pulled off the ranch jobs in the breaking pens, cowboy gangs, fence crews, etc., to bring the raging fire under control. It was a ludicrous way to begin a career as one of the best horsemen and cattlemen in the Islands.

Joe Pacheco trained mules with bossman Kahaikupuna, when all the mules for all the plantations were being supplied by the one ranch where he was employed. He had a long stretch just working with registered stallions, training and exercising horsey aristocrats that went on to their own large shares of fame. Still as a very young man, he took his long turn at breaking horses for the senior cowboys. When only seventeen, he was in the breaking pen as a roughrider, and received wonderful tutelage in 1920 from the aces-high *paniolo*, Willie Palaika.

Later, Pacheco worked with renowned Willie Kaniho, Sr., the bossman of the cowboy gang — and this was like a postgraduate course in super*paniolo* training. In time he became foreman o fthe breaking pen. He went from there to assistant cowboy foreman under Yutaka Kimura, and finally to station manager. Everything he touched became a thing of well-maintained pastures, fine cattle, excellent fences, and neat station premises. He set a great example for every *paniolo* who ever saw him at the craft he knew and loved so well. On a Richter scale of 1 to 10, you'd have to give Joe Pacheco a 9 as a cowman.

XXV

JOHN LEKELESA

A third colorful *paniolo* comes to mind at the moment, and it is John Lekelesa, nicknamed "John Samoa," a long time back. He started at Humuula, just a strong boy with the face and manner of a young friar. It was 1931 when he began, and he sat the saddle for a long 39 years. In those days Humuula was the proving grounds for any and all jobs on the ranch. There was heavy cattle work up there in the heights off Mauna Kea — rough, hungry country — and there were the acres of sheep, fence keeping, maintenance of pipelines, and everything else that goes to keep a vast ranch functioning. He started out at $1.50 a day with board.

Of his 39 years on horseback, he served 34 of them as a top cowboy, the last 25 of which he was in the capacity of foreman. Those days he ran his crew of *paniolos* into amazingly long hours in the saddle, beginning at four in the morning and ending at four in the afternoon — if the job was finished. But even those twelve hours were not the end of labor. With the cattle work completed, the men turned to fence-building or repairing, digging lantana (the range pest), or some other essential work that screamed for attention.

"Those were days of long cattle drives," he explained. "We'd start at one o'clock in the morning to take the cattle to Kawaihae (Hawaii's cattle port), where they were shipped to market on Oahu. In 1960 this was all changed. The cattle are now *trucked* to Kawaihae." Pride was stamped on his strong Hawaiian-Chinese-Samoan face, and he was obviously going back in memory with relish and gusto. He had no regrets.

Testimony to his natural talent as a *paniolo* is found in the fact that he was one of the few men on the ranch who graduated directly into the cowboy gang without first going into the breaking pen for further preparatory schooling and training. Ordinarily this seasoning was mandatory. John Lekelesa retired with the best wishes of everyone connected with the ranch, for he certainly had been a personable and capable *paniolo* throughout the years. His contribution to the Island cattle industry had been substantial.

XXVI

WILLIAM KAWAI

As a fouth *paniolo* who deserves honorable mention, we might put down William Kawai as another outstanding cowboy. Dark, energetic and as honest as a glass of buttermilk, he became a valuable fixture in Big Island cattle raising. For forty-five years he has ridden its ranges and proved his worth time and again as a number-1 man. He was not quite sixteen when the all-time top ranch manager, A.W. Carter, allowed him to ride and work with his regular cowhands on the land's largest ranch. Kawai worked under sharp-eyed Johnny "Poko" Lindsey and Willie Kaniho, Sr., both of whom William gives much credit for his cowboy skill. Today it is still said that "Uncle Bill," as he is now called, has a keen cow sense, and can anticipate what the cattle are going to do before they actually make a move.

He comes of a long line of top-thinking and fast-acting cowmen. His father was William J. Kawai, and his mother was Mary Purdy, the sister of the famous Ikua Purdy. He has been a hard riding and fast roping man right up to December of 1969, when he left the cowboy gang to work at the ranch's Waikii. And there he is very happy as *"papa aipa"* man (overseer of the land). At that place his work is now of a less strenuous nature and he takes pride in still being a cowboy with the added responsibilities of checking paddocks, fences, stock and pipelines in the huge Waikii area.

Too often, men who finally turn their backs on strenuous, hazardous work, find themselves feeling empty — and resigned to the thought that there is nothing left for them but yesterdays. Not Uncle Bill; each new morning, he looks up and thinks, *A handsome sky, a handsome day.* He's still squeezing happiness out of every day and every night. Time never drags its feet for him.

XXVII

FRANK KAWAI

It would be sheer heresy to leave out top *paniolo* Frank Kawai. His years in the saddle and on the business end of a rope have been freighted with color, with excitement, and with genuine productivity. He was the youngest child in the Kawai family of four boys and four girls. Like his brothers, Harry, Duke and David, he decided at an early age to remain on the vast Parker Ranch and work. He would cut out his future there.

In the summer of 1933, while only fourteen years of age, he was taken into the cowboy gang. He was one of three young boys who were allowed this privilege. The other boys were Jiro Yamaguchi and Sonny Taniho. The capable Willie Kaniho, Sr. was their foreman. Their pittance pay of fifty cents a day was more than offset by the honor to ride with some of the best cowboys in all Hawaii. That was worth more to them than all of a rajah's wealth. Riding with legendary *paniolos* was always a moment to clutch, to remember.

During this first summer, Frank's string of horses were handed down from big brother and top cowboy, Harry. Starved for recognition and respect, Frank would have accepted camels. He remembers now that a swift rush of feeling, out of nowhere, caught his throat, and his eyes filled up. That's youth for you...

In 1934, Frank signed on as a full-time ranch hand. His delight knew no bounds. He needed no ticker-tape welcome to the club. Then he was given the privilege to select from the breaking pen sixteen horses for his work string. His brother, Harry, who knew the ranch stock so thoroughly well, was a gravely disappointed man when his kid brother chose a particular filly which was by a stallion named Heinie, to fill out his string of work horses. Worse, he was deeply apprehensive about the selection. Heinie was a trotter, and according to old-timers, was a terror. He was notoriously mean. His foals were tough horses to break and train, but once they were finished they were top performing work horses.

This Heinie horse was both feared and respected by all the *paniolos*. A powerhouse on hooves. The animal protected his band of mares with savage ferociousness. Any horse, with or without a rider, that came into Heinie's pasture or paddock became a target for immediate attack. With all the fury of a wild mountain stallion, Heinie would barge in with ears laid back, teeth bared and hooves lashing out. This packet of a half-ton of screaming horse was something to reckon with, and certainly something hardly worth standing up to.

But Frank Kawai had a way with horses. Every horse he worked with became a polished steed and a worthwhile mount. His golden touch worked miracles with Heinie's get. Those were animals that would have sunk a less vigorous talent than that of Frank's.

After a couple of years with the cowboys, Frank Kawai transferred to the ranch's John Purdy stables where he handled mules exclusively. It was a big operation involving the large, powerful Missouri mules. Breaking those animals was a lot more hairy than breaking wild horses, and most cowboys had a tendency to look the other way when asked to try their hand at it. But Frank was as brave and strong as lions. He did his job in workmanlike fashion — and survived.

In 1940 Frank went into the armed forces, did his bit in the New Hebrides, at Guadalcanal and other bloody points, then was discharged in 1945. He was soon back on the ranch with the Rough-rider Gang. Working with the latter was always a mind-bending adventure. George Purdy was his foreman, and that in itself was like a footballer working with Knute Rockne. Frank worked two years in the breaking pen, then transferred back to the cowboy gang for more years. He is still with the same ranch, rowdy with health, and if he never again throws another rope or rides another range, he can rest on his laurels and with the knowledge that he has made a great lifetime contribution to Hawaii's cattle industry.

XXVIII

GEORGE PURDY

We made slight mention of another *paniolo* who also has cut a solid niche in the annals of Island cattle work. He is robust George Purdy who has been a part of the Waimea cowboy scene for a lifetime. He's a direct descendant of the original Purdy family which started with the colorful hunter-naturalist, William Purdy, whose homestead was located at Pookanaka. George's father was John Purdy, grandson of the famous William, and his mother was the former Annie Ikaia, an extraordinary woman in her own right.

During the summer of 1928, when George was but fifteen, he took his first ranch job picking corn at Waikii. He learned what work was, for one was paid a dollar for a wagonful a day. His group was called the Plaster Gang because picking corn produced painful blisters, and the pickers had to constantly keep adhesive plasters on their sore fingers. After three years of general utility work on the ranch, George Purdy was taken into the breaking pen for the tougher man-maiming chores. He was eighteen by then and had the good fortune of working under the direction of the great horse handler and *paniolo*, Willie Palaika. The experience was a liberal education in that rough department. For lesser men, that work could be a spirit-breaking environment.

Having proved himself the hard way in the breaking pen, George was transferred to the cowboy gang, with the redoubtable Willie Kaniho, Sr. as foreman. Only seasoned first-rate men rode with this outfit, and you either measured up in a hurry — or you lost your spurs to a good man. No one was nursed along with any coddling.

George Purdy passed every rough test, made the grade and remained a cowboy clear up until 1944. He was also one of the *hoau* (cattle shippers) men — a special *paniolo* job requiring specially trained horses. Only a few of the men in the cowboy gang managed to earn this berth, for it required a special technique and one had to be an outstanding *paniolo* to be selected by the foreman for the job. By 1945, George was promoted to foreman of the breaking pen, and he held that important job for a whole twenty years.

George Purdy looked back over his years and explained that a lot of horses were finished during his time. He estimated that between 80 and 90 thoroughbreds and grade horses were broken each year. And that took a lot of doing. To survive all that made a man feel six inches taller and fifty pounds heavier. A rough go for one's money? Well, sometimes it seemed that only a desperate or mentally capsized man would be the one to stay with it . . .

"It used to be," George pointed out, "that a horse was not broken before it was seven years old, nor older than fifteen years of age. A horse was considered fully matured and strong at age seven. It could really be a rough customer for the rough-riders. When the horse was *pau* (through) at the breaking pen, however, it was ready for the cowboys. The horse could handle any hard work required of it without danger of physical strain. It was good for years to come, provided, of course, that it was not abused."

George Purdy was hospitalized in 1968 for several weeks with severe pains in his leg. After he recuperated and returned to work he was assigned to the ranch's race track stables. Twenty-plus years of bucking broncs were more than enough for any man, and the ranch, grateful for his excellent work with horses, transferred George to the track where he could work with young thoroughbred foals. He had surely paid his dues. And there he does a remarkable job today. He is a man who chooses not to go hat in hand to old age.

When asked if he had it to do all over again, would he select the life of a *paniolo*, George answered without equivocation, "Yes, it's a good life."

And you knew by his smile and the timbre of his voice that he meant it. There was no germ of discontent in his answer.

XXIX

DAVID "HOGAN" KAUWE

Another memorable *paniolo* who rates everlasting plaudits is big David "Hogan" Kauwe. He was born on June 2, 1886 on the Parker Ranch, and his place of birth is still marked by tall trees close to the slaughterhouse. The whipcord-strong Kauwe, in his eighties now, still smiles his weather-lined smile when delightedly telling of his hard riding years in North Kahala. And he takes particular pride in relating that he was the last of the famous *paniolo* group of which Tom Lindsey was foreman. He speaks nostalgically of famous individuals like William K. Lindsey and the other members of that clan like Bill and George. He waxes positively maudlin when referring to the likes of Kaihue Pua'o, Willie Spencer, Archie Kaawa, John Chesebro, Kaiipapau Kaliikini, and Kaliko Mainaaupo.

"We worked together," he explains, letting memory run way back. "And we had fun together. You don't find these things nowadays; the way those cowboys stuck together, helped each other. Any trouble, we were all together. It's not that way today, seems to me."

His solid six feet of hard-knit body is mute testimony to his lifetime of tending cattle, chasing and roping cattle, wrestling cattle, breaking horses, and all the many other things which fall to the *paniolo's* lot. His body, ancient though it is now, is symbolic of the good rugged life in the saddle, the exposure to wind, sun and rain, and the plain, substantial fare of the cowpoke. The usual cruelty of time had never really gotten to him.

"I was a school dropout at thirteen," he said. "But I was big for my age and could do a man's job, so I turned to pick and shovel work for a living. Then I went to work in 1899 for Tom Lindsey's cowboy gang on the Parker Ranch."

He grinned in retrospect, adding, "Manager Alfred Carter switched me to the roughrider gang. Barney Judd and I were the first of that famous group. I remember the stable was where the ranch saddlery is now. I broke eight horses there to begin with."

But a little strain of sadness crept into his voice when he described how he was fired when a horse died under him on the tough Waipio trail. "Mr. Carter didn't like it — and canned me. B-but I still believe that horse had heart trouble."

It had left him jobless for a long, lonely time and he felt he had got only a 50 per cent shake of the dice from life. But even at eighteen — what with all the bounce and buoyancy of youth — it didn't deter him from wooing and winning the hand of Rena Mainaaupo, the daughter of Kalipo Mainaaupo, grandfather of "Billy Boy" Lindsey. Hogan had been roping with the legendary *paniolo*, Ikua Purdy, when he first met the girl, had learned his cowman's lessons well, and had not believed that unemployment could descend upon him. He was a prideful young man at his trade; any other field was second best. He brooded.

In thinking back, he tried to recall the ache of those days of unemployment; it seemed to stick in the shadows of his mind. Finally, he explained, that as a last resort, he went to work on the mule train which was then supplying Hamakua ditch, but he longed for the regulation *paniolo* work which was in him like a heart throb.

After an interim of doing this secondary work and sweating out the tedium of it, he ultimately worked up the nerve to beard the strict disciplinarian, Carter, in his office den. He pleaded for reinstatement. It must be understood that Carter was bossman of scores of *paniolos*, and he was leaving a large mark on the Hawaiian cattle industry. He was no common man. In any event, the ranch manager relented and put Hogan to work cutting *mamani* fence posts with Hogan's brother, Puna. Hogan toiled hard at a lot of other mundane jobs that year, still hurting and brooding over not being in the role of a *paniolo*. He felt lonely beyond the telling.

Then, before too many months slogged by, Mr. Carter again allowed him back on the *paniolo* gang. Hogan was reborn. Hogan had not been searching for any sinecure in the way of a job; he went back to cowboying with flags flying, knowing full well what the hazards would be and what heavy labor would be required. His happiness was now boundless, and his face showed everything — everything.

Like all *paniolos*, Hogan had his share of accidents, his quota of shatteringly bad moments — but also his allotment of joys. Meanwhile, he was pitiless with his own energy. At Puukaaliali he fell with his horse and broke his collarbone and was laid up for over two months. While riding alone in 1936, he was thrown from his fractious mount near Pauwanui Paddock and badly fractured his right ankle. When his *paniolo* pals found his riderless horse in the ranch corral, they turned instantly to ransacking the valleys, hills and ravines for him. After much combing of rugged terrain, they found him and his breathing was like the sound of cloth being torn.

Carefully, they brought the injured man back to the ranch for medical attention. His injuries that time necessitated his being laid up for a full sixteen months, and he was to feel the pain of the crushing fall right up to this day. The accident had diminished him, there was no blinking it.

But, despite injuries of various kinds, lethal and crippling, Hogan did not go into retirement until 1957. He needed another tough ride like Custer needed another Indian. He had given a full fifty years of faithful service. He's still fighting the grave to a standstill.

Fondly and nostalgically, he says, "Those were happy years. Sometimes I was acting foreman and assistant foreman. When John lindsey retired, Willie Kaniho was brought down as foreman from Humuula." Here was a man who had been touched by those lonesome hills and valleys of Hawaii and all the island things that made the memory sweet-smelling like a load of hay.

Even accidents didn't lessen Hogan's love of his work. These were all memoried things straight out of David Hogan Kauwe's heart, for he was looking back at those long-ago events — most of them, anyhow — with the keenest of relish. And with a deep, wistful sentiment.

XXX

WILLIE KANIHO

Hogan Kauwe's reference to Willie Kaniho brings us to another brief *paniolo* profile. He, too, is a 50-year cowboy — a multitalented and much loved man. On July 4, 1964, the Parker Ranch held a "Willie Kaniho Day," and it was as sober as it was festive; sober for those who realized that they were losing a buddy who had worked at their side for a full half-century; festive because the ranch had gone all-out in honoring this retiring *paniolo*. Hundreds and hundreds of people were on hand to pay tribute to the gray, but indestructible-looking, man who still rode his horse like a crusader.

Willie Kaniho's indestructibility stood him in good stead again this day because, paradoxically, he had to fill the dual role of director of the whole show, plus being the guest of honor. That's a good trick when one can do it; and he did it with all the aplomb and efficiency with which he always did everything.

Willie's mood matched the weather. It was the kind of day when, on the vast prairie and range there is a brooding loneliness of the land, like the hazed blue of the flat sky, hanging in silence over everything. On the long undulating slopes of Mauna Kea sunshine was lying like a golden shawl across their shoulders. Everything was strangely sweet-sad, and Willie was glad the tradewinds were blowing in with unnameable hope, promise and cheer — for this day was his swan song; or close to it.

Above the festivities flew a banner carrying the honored guest's name, and below it were the Fourth of July events which included fancy roping and riding, martial music from the U.S. Army band from Schofield Barracks of Oahu, speeches, and all the other holiday activities that celebrate such an occasion.

Addressed as "Willie" by everyone, Kaniho became something of an institution among the *paniolos* and cattlemen at large. He was born at Kalopa on the Big Island on March 18, 1894. His father worked on the ranch at Humuula. His mother's maiden name was Kanaihola Papa, and her parents also worked at Humuula. Due to the death of William Kaniho, Sr. shortly after Willie was born, the grandparents — the John Papas — assumed the raising of the fatherless child. Willie left school at age fourteen and also went to work at Humuula.

Later, for reasons of obtaining better pay, Willie, Foreman Ikua Purdy, and a third *paniolo*, Kainapau Kailikini, left to ride for a Kauai ranch. Willie's hiatus from the Big Island ranch didn't last too long, for Manager Alfred Carter sought him out and rehired him as a foreman. At Humuula, Willie replaced a good man, Junichi Ishizu, who had been killed in a flash flood when he and his horse had been caught in the torrent. Willie went on to put in a full half century riding for the ranch of his first love.

He's out to pasture now, and in the sunset of his years, but he still revels in speaking fondly of the *paniolos* with whom he rode. He likes to jawbone quietly about his six years at Humuula, returning to the cowboy gang to replace Johnny Lindsey, Sr., when the latter retired. With particular relish he talks of his eight years as *paniolo* foreman in the days when the cattle were lashed to longboats in the surf, then floated out to the steamers and lifted aboard for further transfer. That was a very special era to him.

In 1959 he was brought into Waimea as general foreman under Manager Richard Penhallow. His responsibilities mounted along with his pay and his increased know-how, and he wonders now how he managed to have any homelife at all. On the other hand, he laughs and points with pride at the fact that he has ten children and thirty grandchildren. Widowed twice and married for the third time, he, uh, yes — he somehow worked a little homelife into those later years...

But time gallops on — or *tempus fugits,* as it has been said — and on Octoboer 18, 1966, Willie Kaniho finally hung up his spurs, turned his work horse loose for the last time, and accepted retirement with fifty years of *panioloing* back of him. He's almost as pompous as a parrot over his son, Daniel, stepping into his shoes on the ranch. To quote him, he says, "Hawaii cattlework has been good to us, so we'd damned-well put something back."

Willie Kaniho is coasting now and there's an ebb tide look in his face, but he still rejoices in talking fondly of the *paniolos* with whom he worked. He delights in reminiscing about Kaliko Mainaaupo, Tom and Albert Lindsey, Frank Vierra, Alex Akau, Sr., William Campbell, Hogan Kauwe, Awili Lanakila, Harry Kawai, Joe Pacheco, John Lekelesa, and other members of that legendary *paniolo* crew. They were all stout men, and Willie Kaniho was one of the stoutest.

XXXI

HARRY KAWAI

The family name of Kawai draws plenty of water in *paniolo* annals. And none more than that of Harry Kawai. No more colorful *paniolo* than this very Harry ever rode the North and South Kohala ranges. He closed out a fifty-year hitch on the Big Isle on November 30, 1968. By his own definition, riding its huge ranges and working with its livestock was his biggest satisfaction in life. He is dark, brooding, and an ax handle across the shoulders. Well-knit is a good description of his frame. The mark of the open air and wide spaces is upon him.

He was born the son of Dick Kawai and Likolehua Spencer Kawai. It's understandable that he was a *paniolo* to the manner born, for his own father had put in his own lifetime there riding the same land. So, with ranching already coursing in his veins, Harry started out as a boy doing all the minor and mundane jobs on the ranch such as cutting corn suckers, digging guava and chopping out lantana. It was all part of the indoctrination whereby a young man learns his ranching.

Of course the jobs got bigger, tougher and packed with more responsibility. But as tough and poor paying as they were, Harry Kawai never lost the awareness that he was working and living in an incredibly beautiful country, and, best of all, working with men who liked and respected him. It was the old story of when you're in love everything seems interesting. Well, Harry was in love with Hawaii and cattlework. The regular *paniolo* gang was made up of such a select and capable brand of men that only the most outstanding of ropers and horsemen could possibly hack it.

Several times Harry came close to being called to duty on the gang, but, as he laughed in retrospect, "Close counts only in horseshoe pitchin'." To him that *paniolo* gang began to appear as a "closed corporation." Striving for it, and not making it, began to give him a walkabout restlessness. When asked why he wasn't working with the *paniolo* gang, he used to look grave, start to make up some alibis for his melancholia, then would abandon them all and resign himself to letting time take its course. He buried his dreams in silence. Worse, trying to figure out how to get on the gang was like trying to sort out a bag of snakes.

Then, finally, in August of 1925 the death of Iakopa Kahaikupuna left a vacancy — an almost unfillable vacancy. That was some vacuum to fill, for Iakopa was known as the "Will Rogers" of Hawaii — a versatile, all-around man with horse, rope and steer. But Harry Kawai had proved big, strong and talented himself; a man for all seasons. He had worked along with and studied the likes of Iakopa on the ranch, and had soaked up much of their knowhow and skill.

He met the challenge as a substitute and for seventeen years he punched cattle, shipped cattle, and trained shipping horses for this specialized work. He was a Trojan for work, and as brave as a badger. In no time he was being likened to some of the all-time greats among the *paniolos*.

Harry worked at his robust, but dangerous, trade without serious mishap until just before the outbreak of World War II. Then it was that he suffered a ruptured appendix that sent him in a rush to the Kohala Hospital.

Quick, emergency surgery saved his life. It had been a close call, and for awhile he had felt himself slipping across the honed edge of death.

But his indomitable will and ironclad constitution kept him from being grounded too long. He ultimately came back to his *paniolo* chores like a wholly revived champion fighter. In fact, with his natural strength fully returned, he was assigned to the breaking pen as foreman. He worked right along with his men on this grueling, dangerous operation. Depending upon how fractious and wild the animals were, each broncobuster broke from five to ten horses a day — and every ride could be a *paniolo's* last.

Those were the days when so many of the better ranch horses were being purchased by the United States Army, and good, stout horseflesh was at a premium. Foremanship of the breaking pen for Harry continued up into 1947 when his boss, Alfred Carter, moved him up to the position as general foreman of the whole vast ranch. Due to this being such a ponderous ranch operation, this was a job fraught with much responsibility and requiring a man with rare *paniolo* skill. Harry Kawai was that man; he had the knowledge, imagination and talent to hold the job successfully for twenty-one years. Carter's judgement of the man had been righter than Solomon.

But the rigors of a *paniolo's* life, where Harry spared himself not at all, began to take something of a toll. After his appendix operation, he was subjected to eleven major surgeries. They just about did him in, and he began to feel that he was coming unglued. His clothes hung on him as though they were still in the closet. There was a late-afternoon look in his eyes, and also his color was not good. More and more he seemed like a man who is listening to two conversations at once. In short, with all his hospitalizations his body had begun to pay at usurious rates, and the bedridden sieges robbed him of much of his clout.

So, in November of 1968, with sixty years on his back, Harry elected to do his last day of work on the job. It was a hard choice for him because he still loved working the land; it was as important to him as his next breath. The ranch and its operation had been a live thing with him, and he liked to breathe along with it. It's understandable, then, that sadness lay heavily upon him when he finally decided to close the door on his active life. The very prospects of it had touched a nerve.

The final day he went out onto the range for the last time to check the water lines and range paddocks. To him he felt it was like a last roundup, and his eyes were wells of sorrow. One word from him about this being his last sally out onto the open spots, and he would have had many friends accompanying him for auld lang syne. But he chose to be silent about it and not feel compelled to say a lot of melancholy farewells. Although the ranch had been the breath of life to him, maudlin he would not be; not in anyone's presence, at least. Too, he felt it would make it easier on his co-workers.

He took that last ride by himself — a brooding trip because he was so full of the things that had been. A soothing wind drifted from the Mauna Kea slopes and cooled the lava wastes. He turned his head seaward and knew that way beyond him there was nothing but the solitary shoreline in that direction, gleaming sea and hanging sky. He road on inland and telling himself that he was through put a tight knot in his stomach as though he'd just been told he had terminal cancer. Throughout the ride, he was totally caught up in the web of the past, and each time he said good-bye to a remembered place, it seemed that a million violins began to cry.

When Harry returned to ranch headquarters, he trudged up the steps like a man with iron shoes. He came into the office and there were those who heard him say simply: "Masao Yoshimatsu, Frank Vierra, and I are all that are left of the 1918 Opae Gang." He was very quiet about it and it was as though he spoke only to himself. It was obvious to those within earshot that Harry felt he was a man who had run his race. Harry looked around at the ranch owner and several others close by. "I want to thank Richard Smart and all of the ranch hands for making my job a pleasant one. God bless you all."

It was that simple a parting. Harry Kawai went into no jeremiad, spoke no recrimination, mentioned no regrets, sought no sympathy or praise. He had had a good life — and was grateful.

XXXII

A *PANIOLO* SALUTE

So many top *paniolos* have contributed to making history in Island cattle ranching that it seems criminal not to do profiles on all of them. But the limited space here gives us room for only a few in these pages. It doesn't, for an instant, mean that the unmentioned ones have been lacking in talent, dedication, color or contribution.

In short, these profiles portray but a handful of the *paniolos* who have done such splendid work in their chosen profession. They deserve everlasting salutes for their contributions to the building of Hawaii's cattle industry and the general economy of the Fiftieth State. So many of them have been champions in their own right, men to be remembered.

Well-done, gentlemen of the range — an honest low bow to you and your fine women who have backed you so faithfully and steadfastly. To all of you — *Mahalo nui nui!* (a big thank you).